Presented by

RAMPARTS MAGAZINE

Learning Resources Center

SOUTHEASTERN COMMUNITY COLLEGE
Whiteville, North Carolina 28472

THE
POLITICS OF ECOLOGY

Books by James Ridgeway

THE CLOSED CORPORATION
THE POLITICS OF ECOLOGY

THE
POLITICS OF ECOLOGY
James Ridgeway

E. P. DUTTON & CO., INC.   NEW YORK   1970

This book is manufactured using reclaimed paper fiber. The paper
is Warren's Sludge Opaque, Basis 60, Antique finish. S. D. Warren
Company, a division of Scott Paper Company, manufactured this
paper using 50 percent reclaimed effluent from the primary clarifier
of their mill in Westbrook, Maine.
Several years ago this material polluted a river. In recent years it
has been recovered and used as land fill. S. D. Warren Company states
that this is one of the first uses of paper-mill sludge in the manufacture of book paper.

For My Mother and Father

Acknowledgments

I am especially grateful to Frances Lang for her help in preparing this book. She did much of the historical research as well as a great deal of reporting. Also I wish to thank my friends at *Hard Times*, Robb Burlage, Bettina Conner, my wife Pat; and the various employees of the Federal government who provided useful advice and information. I am grateful to Hal Scharlatt for his patience and friendship.

Contents

THE
POLITICS OF ECOLOGY

1 | Earth Day

Ecology became a popular issue during the early spring of 1970 because it momentarily offered the prospect of a new politics, a new set of symbols with which to rework the social order.

Ecology offered liberal-minded people what they had longed for, a safe, rational and above all peaceful way of seeming to re-make society, limiting the growth of capitalism, preserving the natural resources through pollution control, developing a more coherent central state; in short, establishing programs and plans for correcting the flaws in what many perceived to be a fundamentally reliable, sound political system.

Advanced and progressive industrialists shared this perception. Since the beginning of the century, large corporations have often argued for national conservation policies. Beginning with the administration of Theodore Roosevelt, they entered into a partnership with government in managing those resources. The ecology issue offered them a chance to re-assert their primacy in that coalition.

Ecology touched millions of ordinary white middle-class people as no other political issue had for years. It aroused smouldering passions for a pre-industrial past, reviving the tradition of American individualism, of man in nature. But then, ecology also affirmed our faith in science, in the steady march of technology which could solve the problems of pollution.

For the Nixon government, stumbling amidst the wreckage of its own ambitions, ecology offered a domestic program that would seemingly divert attention away from Southeast Asia.

But the protestations of a President, the celebrations of Earth Day, a crusade by students with their arms full of non-returnable bottles, could not obscure the Indochina war, the organized political repression, economic disintegration, dissolution of the central government; all those signs of a society never before so split asunder, perched on the edge of ruinous race and class war.

Once the hysteria of the moment had passed, the politics of ecology seemed altogether dull, complicated and in the end paralyzing, bestowing on the participants a special sense of futility and alienation. It was an issue which told us only that we are all victims and that nothing changes.

Nonetheless, that exercise in futility had its point. It provided a cover behind which the ecology interests could wage their struggle for control of natural resources. This book sets out to explore that struggle, describing the underground war for control of the water-pollution programs, the key to control of other environmental policies, and the battle among the petroleum trusts for domination of the world energy markets. For our purposes, ecology embraces "pollution" and "conservation," words used at other times and in other contexts to describe the political economy's natural resource policy.

Environmental pollution, as we know it, first appeared in the English towns during the height of the Industrial Revolution more than 150 years ago. The ideas for environmental controls and the technology to put the ideas into practice were invented during that period by the Benthamites as a means of advancing the interests of the manufacturing class. Sanitation, as originally conceived by the English reformers, provided a way to order and organize the work force so as to increase its productivity. From the beginning, the idea of pollution control was not to restrict industrial development, but to ensure it.

Virtually all our modern programs for controlling pollution begin with the premise that controlling pollution is essential for industrial growth. The technology is little changed since the mid-nineteenth century, and many towns and cities in the United States of 1970 do not possess the rudimentary equipment for treating sewage recommended in 1900. Few of the political reforms enacted within the past few years go so far as the methods argued for by the Benthamites one hundred years ago. Our efforts have resulted in pouring money into the construction of elaborate, practically useless sewage systems, and utterly meaningless programs for controlling air pollution. Billions of dollars have done little more than keep construction workers and civilian engineers employed at the same time ensuring fiefdoms for quarreling sects of bureaucrats.

The water and air-pollution reforms of the last ten years have not halted, or much slowed, the rate of pollution. On the contrary they hold out the prospect of worsening the situation.

Meanwhile, the real struggle for power goes on in a war among the giant petroleum companies for control of energy sources. The ecologists, ideologues for the ecology crusade, ironically function as a cover for the energy game. They talk in radical terms about reorganizing society, about population control, but they merely serve to distract attention from the central issue.

Burning coal, gas or oil creates energy, and creates most of the pollution. At first glance it may seem surprising, but companies which create most of the energy and cause the pollution are the leaders in the anti-pollution crusade. These large corporations anticipate that by dominating the ecology movement, they can influence the rate and manner in which pollution control is achieved. In some instances the companies own subsidiaries which can sell products to control the pollution which they create through other subsidiaries. Most important, the ecology crusade serves as a screen behind which they are amassing control over different resources. (To cite but one example, Standard Oil of New Jersey, the mother of trusts and greatest oil company in the world, moving beneath the ecology banner, has stealthily gained a near corner on coal resources in America.) When that job is complete, a handful of international corporations will control the different fuels, and be in a position to dictate world-wide uses of energy and, hence, determine the rate of pollution.

2 | Chadwick's Inquiry

Cholera first appeared in Europe in 1832 and rapidly spread across the continent. A devastating plague settled upon England, killing thousands of people who inhabited the squalid courts and back alleys of the crowded towns. Cholera reappeared in 1848 and again in 1853. The presence of the disease helped hasten major health reforms. Those reforms, which began at mid-century, created a centralized sanitary and health system.

While the health reforms served a humanitarian purpose, they were also meant to serve the more utilitarian ends of the Benthamites, who believed disease was retarding the growth of the manufacturing class. Improvement of the sewage system and water supply were viewed as ways of promoting the growth of modern capitalism.

The laboring people who packed into towns during the nineteenth century lived in abominable conditions. The drinking water was foul, and drainage for sewage and garbage was often nonexistent. Of fifty large towns, only six had a good supply of water; thirteen were indifferent and thirty-one had impure water. In London, where nine companies controlled the water supplies, three quarters of a million poor people begged or stole their drinking water for want of money to buy it. People hauled water for miles, oftentimes collecting it from ditches. The situation in Tranent, a Scottish mining town, was typical: "When cholera prevailed in that district, some of the patients suffered very much indeed from want of water, and that so great was the privation, that on that calamitous occasion people went into the ploughed fields and gathered rain water which collected in depressions in the ground and actually in the prints made by horses' feet." Ten wells provided Tranent with water, but much of the water was diverted from the wells to a coal pit on the outskirts of the town, and in times of drought the water was reduced to a trickle. "On these occasions of scarcity great crowds of women and children assemble at these places (the pumps), waiting their 'turn' as it is termed. I have seen women fighting

for water. The wells are sometimes frequented the whole
night."

Few dwellings were connected to sewers, even where
sewers existed. In some parts of the country it was against
the law to connect house drains to sewers. The sewers
were reserved for storm drainage. In some instances, one
privy would serve thirty buildings. In Manchester in
1843, there were thirty-three privies for 7,000 people.
Even where an advanced sense of sanitation had led to
construction of sewers, they would end abruptly, dis-
pensing their contents into the middle of a street or
lane.

Cesspools were usually located adjacent to the tene-
ments, and the contents seeped through the walls of ad-
joining buildings. In Liverpool, where the most horrid
conditions existed, 40,000 people lived in dank cellars.
Dr. Duncan, a local doctor, wrote in 1842: "From the
absence of drains and sewers, there are of course few
cellars entirely free from damp; many of those in low
situations are literally inundated after a fall of rain.
To remedy the evil, the inhabitants frequently make
little holes or wells at the foot of the cellar steps or in
the floor itself; and notwithstanding these contrivances,
it has been necessary in some cases to take the door off its
hinges and lay it on the floor supported by bricks, in
order to protect the inhabitants from the wet. Nor is
this the full extent of the evil; the fluid matter of the
court privies sometimes oozes through into the adjoin-
ing cellars rendering them uninhabitable by anyone

whose olefactories retain the slightest sensibility. In one cellar in Lace-street I was told that the filthy water then collected measured not less than two feet in depth; and in another cellar, a well, four feet deep, into which this stinking fluid was allowed to drain, was discovered below the bed where the family slept!"

Even where a person could afford to hire a carter to dig out a cesspool and haul off the contents, it was far from satisfactory. The carter usually did his business at night, and in all likelihood he would haul the filth to the nearest open space or by-lane and dump it there, rather than take the trouble of carting it out of London into the country. Some of the modern buildings in London were connected to sewers, and they had water closets. The city sewers, however, were not regularly flushed out. They were flat-bottomed and flat-sided affairs. Sewage would collect and back up into the houses; gases exploded. When a sewer was clogged, there was nothing for it but to rip open the street, and amidst a foul stench, send down a gang of sewermen to shovel the filth. Buckets were then hoisted to the street by windlass, and dumped onto the sidewalks for drying. Sewer gangs feared for their lives from the exploding gases, and emerged from the task pale, drawn and feverish.

Cholera and typhus took their worst toll among the working-class districts. In Liverpool in 1840, where conditions were at their very worst, one in twenty-five people were attacked by fever. Death rates among the laboring population were staggering:

LIVERPOOL, 1840

	Average age of the
Number of deaths	deceased
137 Gentry and professional persons, etc.	35 years
1,738 Tradesmen and their families	22 years
5,597 Laborers, mechanics and servants, etc.	15 years

Of the deaths among the laboring population in Liverpool, no less than 62 percent were children under five years old.

In 1839, the Poor Law Commission ordered an "Inquiry into the Sanitary Condition of the Labouring Population of Great Britain." The investigation was led by Sir Edwin Chadwick, a disciple of Bentham. Chadwick was a passionate reformer. As secretary of the Commission he had already proved highly unpopular for his utilitarian workhouse scheme to instruct the idle poor in the virtues of hard work by denying them relief unless they entered a workhouse. There, according to theory, they might be supervised in useful employment. Chadwick believed there was plenty of work available if the poor would only seek it out. However, there was not enough work and the workhouses turned out to be prisons where the poor spent their time picking bones to get enough to eat.

Chadwick's inquiry on sanitation, presented to Parliament in 1842, included reports on various aspects of health and sanitation, ranging from statistical studies of diseased districts to commentaries on new types of sewers. Doctors reported on local conditions throughout the country; surveys were made of sewage works on the continent, and Chadwick himself toured England and Scotland inquiring into everything from sewage manure to working conditions in factories and mines. Chadwick used the report to push for health reform and to further the drive for growth of a centralized government. Driven by utilitarian theory, consumed with the mechanics of government, he typified the new class of bureaucrat.

Chadwick was convinced sewage was wasted, and could be put to some efficient and economically beneficial use. He was impressed with experiments in Edinburgh: "A practical example of the money value which lies in the refuse of a town, when removed in the cheapest manner, and applied in the form best adapted to production, viz., by a system of cleansing by water, is afforded in connection with the city of Edinburgh. In the course of the sanitary inquiry in that city the particular attention of Dr. Arnott and myself was directed to the effects of some offensive irrigation of the land which had taken place in the immediate vicinity of that city. It appears that the contents of a large proportion of the sinks, drains and privies of that city are conveyed in covered sewers to the eastern suburb of the town, where they are emptied into a stream called the Foul Burn, which passes ultimately into the sea. The stream is thus made into a large uncovered sewer or drain. Several years ago some of the occupiers of the land in the immediate vicinity of this stream diverted parts of it, and collected the soil which it contained in tanks for use as manure. After this practice had been adopted for a long period, the farmers in the vicinity gradually found that the most beneficial mode of applying the manure was in liquid form, and they conducted the stream over their meadows by irrigation. Others, perceiving the extraordinary fertility thus obtained, followed the example, and by degrees about 300 acres of meadow, chiefly in the eastern parts of that city, but all in its immediate vicinity, and the greater part of it in the neighborhood of the palace of Holyrood, have been systematically irrigated with the contents of this common sewer. From some of this land so irrigated, four or five crops a year have been obtained; land once worth from 40s. to 50s. per acre now lets for very high sums. . . ."

The sewage was in such demand that owners and tenants of lands adjacent to the Foul Burn claimed they had a legal right to the city's manure and they defended the practice in court. They claimed that proposals to divert the sewers would deprive the city of Edinburgh of milk and butter yielded by cows which fed on grasses irrigated by the sewers.

Chadwick was dubious about employing these methods so near the densely populated center of the city, but he was very much interested in the basic procedures. He argued that the proper course would be to drain the cesspools of Edinburgh into covered sewers, flush the contents of the sewers out onto the countryside, away from the town, where it could be used for manure. He believed such a scheme could be a real economic benefit in London, where sewage was worth double the price then paid for water, and it was being wasted by being dumped into the Thames or the streets.

Chadwick was among the first advocates of a combined water supply and sewage-disposal system. He believed a system of underground pipes could carry drinking water under pressure to the house, then transport sewage diluted in water out to the countryside, where it could serve some use in irrigation. That same system could provide water for cleaning streets and fighting fires.

The sanitary report reflected Chadwick's interest in parliamentary reform and his belief in the necessity of expanding central government. That reform, according to utilitarian principles, would further the interests of the manufacturers and help maintain order among the poor. He believed the capitalists were denied the benefit of workers' production because of conditions in which they lived. And he argued for changes which would result in increased production by lengthening workers' lives.

Moreover, sanitation could be a method of organizing the poor. Older workers were a temporizing influence, not much given to running in mobs and taking part in trade-union activities. The moral habits of the poor would improve if their neighborhood were cleansed. He noted there was little alcoholism in clean, poor districts. As an example of the sort of improvement which came from proper sanitation practices, Chadwick approvingly

quotes Villermé, the reporter of a committee of the
Royal Academy of Medicine at Paris, who had written:
"At Varregio in the principality of Lucca, the inhabitants, few in number, barbarous, and miserable, were
annually, from time immemorial, attacked about the
same period with agues; but in 1741 floodgates were constructed which permitted the escape into the sea of
waters from the marshes, preventing at the same time
the ingress of the ocean to these marches both from tides
and storms. This contrivance, which permanently sup-
pressed the marsh, also expelled the fevers. In short, the
canton of Varregio is at the present day one of the
healthiest, most industrious, and richest on the coast of
Tuscany; and a part of those families whose boorish
ancestors sunk under the epidemics of the *aria cativa,*
without knowledge to protect themselves, enjoy a health,
a vigour, a longevity, and a moral character unknown to
their ancestors."

The sanitary report called for establishing a central
health board which could plan water supply and sewage-
disposal systems and then require installation in localities if the officials did not act on their own. The measure
was opposed by local property holders, who did not
want to pay more in taxes and by the iron and coke
manufacturers, who particularly resisted a "smoke sup-
pression" clause in the proposal. In its final form the
Public Health Act of 1848 established the principle of
centralized control over public health, but lacked any
real authority to carry out the reform. The act created a

three-member public-health board, of which Chadwick became the leader.

For the next five years Chadwick fought a bitter, often futile battle to improve sanitation in England. He was opposed on all sides. Local sewer authorities, who had held that sewers were meant for storm waters and not as conduits for domestic wastes, resisted his efforts. Poor people who lived in the worst districts were suspicious of the modern scientific attacks on disease. During the 1832 cholera epidemic, rumor spread that the government had deliberately poisoned the water supplies to reduce the population. For their part, sewer authorities opposed flushing out clogged sewers for fear the district would get a bad name as a center of disease.

In 1848, cholera reappeared. In 1832, 16,000 people died of the disease, by 1849 the number reached 72,000. Nonetheless, towns sewered by the act of 1848 had a lower death rate. As the cholera epidemic spread, the health board sent teams to severely diseased areas, and ordered streets cleaned and buildings whitewashed.

The health board repeatedly sought to establish a central water-supply system for London, but the different private water companies fought off every move in that direction. The Secretary of War, who owned shares in a water company, fought it in Parliament. The solicitor of the Treasury, who also served as a water-company director, was against the idea. Clerks in the House of Commons held stock in water companies. The powerful in London were so thoroughly intertwined with the

greedy, inefficient water companies that all efforts to
establish a centralized system which could reach all the
people were resisted until 1902, when the water companies were consolidated.

For the next fifty years a series of parliamentary committees and Royal commissions investigated the different
aspects of pollution, and all of the investigations resulted
in recommendations for purifying sewage by spreading
it on the land. While sewage farming might have some
utility, the main reason was to avoid polluting the
streams and rivers, and thereby spreading disease into
the water supply. A Royal Commission in 1861 declared:
"In our former report we stated and we have since seen
nothing which could induce us to any degree to alter
our opinion that to obtain the greatest amount of good
from the utilization of sewage as manure, concurrently
with the fullest immunity from the evils arising from
its discharge into rivers as a noxious and pestilential
liquid, to obtain the good and avoid the evil, one only
thoroughly efficient mode of treatment could be pre-
scribed—that, namely of its direct application to the land
in the liquid form." A Select Committee in 1864 and a
parliamentary report in 1865 both urged a sanitation
system where sewage was spread on the land. The Rivers
Pollution Commission in 1868 investigated the possibility of using chemicals to treat sewage; they were
unimpressed and agreed the most sensible solution was
to spread the sewage on the land. Three years later a

commission recommended legislation which would give towns the right to seize land for use in purifying sewage. In 1876 the Rivers Pollution Act made it an offense to discharge sewage into a stream.

But enforcement proceedings under the Rivers Pollution Act could not be initiated for two months after the alleged offender was given written notice. The court could then send around "skilled parties" to advise the offending party on the measures he might undertake to make the sewage harmless. Special exceptions were made for manufacturers. Before a sanitary authority could require a manufacturer to clean up his sewage, it needed to obtain permission from the local governing board. The local board was required to hold a hearing, and in giving consent to any enforcement proceeding "must have regard to the industrial interests involved in the case and to the circumstances and requirements of the locality." And the governing board ought not give its consent to proceedings by a sanitary authority of any district which is the seat of any manufacturing industry unless it is satisfied that the "means for rendering harmless the poisonous, noxious, or polluting liquids proceeding from the processes of such manufacturers are reasonably practicable . . . and that no material injury will be inflicted by such proceedings on the interests of such industry."

In the end there was very little to be done about industrial pollution except to plead with manufacturers. If the manufacturer said he was doing all that could be done, that was the end of it. The general feeling of the time was expressed by a witness before the Royal Commission on Sewage Disposal in 1901: "You have to deal with this question from a common sense point of view. What is the object we have in view? The object is to make rivers as little of a nuisance as we possibly can. But that must be consistent with supporting industries of the neighborhood. If you go on to insist on conditions which it is impossible to carry out except to the detriment and destruction of those industries, then I say you are doing far more harm than good."

By the end of the century, the Royal Commission on Sewage Disposal began to investigate the idea of establishing water-quality standards. But witnesses before the Commission argued that this was not a sensible course, for by establishing standards, one was in fact legitimizing pollution at a certain level. If the standards were set too high, then the corporations would say it was impossible for them to comply. The enforcement procedures under the Rivers Pollution Act was slow. A company could get by for two years before submitting a plan for controlling pollution. It was impossible to treat industrial wastes, because they upset the treatment process. A witness said, "In many instances the sewage works were closed at night, the men in charge left them shut down, and turned all the sewage into the streams by the storm water overflow."

The Commission heard arguments to the effect it was not necessary to fine polluters. It was argued that in many rivers, into which pollutants were discharged, it made little difference because they were diluted due to the volume of the water. Although irrigation by land was the best way for handling sewage, there was not enough land available.

In its conclusions, the British Commissions of the early 1900s gave up on the idea of treating sewage by spreading it on the land, and adopted a version of what has come to be known as the "dilution" or "assimilation" theory. Under that scheme, sewage is treated, by various chemical or biological processes, and then discharged into the stream in quantities which the stream is judged to be capable of diluting. The stream is formally viewed as an integral part of the sewage treatment process. Then, as now, the dilution theory is the chief rationale for water pollution.

In the last decade of the nineteenth century pollution became a serious problem in the United States, especially in New England, where the mill towns were beset by conditions similar to those in England a half century earlier. The towns were usually located on small, low flowing rivers, where drinking water and sewage readily combined, and, as a result, disease rates were high. In 1872, the Massachusetts State Board of Health warned of pollution levels equalling those in England, and in 1886 the Board of Health was given authority over all inland waters in the state as part of a plan to stop pollution. Towns and cities were required to submit plans for sewage disposal and drinking-water supplies to the board. So that it might keep better informed on these

matters, the state put up money to finance research into
pollution. That led to the creation of the Lawrence Experiment Station at Lawrence, Massachusetts, on the
Merrimack River. The Merrimack, then as now, was
badly polluted, and the town of Lawrence was plagued
with epidemics of typhoid. The first director of the
station was Hiram Francis Mills, a hydraulic engineer,
who had worked for the company which ran the dam on
the Merrimack. Mills hired assistants from MIT and
began to assemble the first of a small group of sanitary
engineers. These men traveled to Europe and inspected
sewers and research experiments, especially in England
and Germany. They were aware of the work on the
Ruhr, where the engineer Karl Imhoff was building
settling tanks for filtering sewage. They kept abreast of
the different commissions on sewage in England.

Massachusetts discarded the idea of spreading sewage
on the land because of the enormous amount of space
required. One of the health department's early reports
points out that in England an average of 4,000 gallons of
sewage could be purified per acre per day. In Germany,
the rate was a bit less. In Paris, it was much higher.
(There, sewage was poured over loose sand to irrigate
cabbages, and 11,000 gallons per acre could be handled
without drowning the crops.) But to attempt to purify
the sewage from a town such as Lawrence in this manner
would have required land equal to a quarter of the size
of the whole city. And if the land were not suitable, the
sewage might lie across it in a thick, putrifying mat.

In England, the concept of irrigation had given way

to filtration. Sewage was passed through earthy materials of one sort or another, usually sand or gravel, without any attempt being made to raise crops. In 1868, at Ealing, sewage was distributed over burnt clay and coal. Different experiments were made in Europe in an effort to figure out what types of soil had a purifying effect. In France, Schoesling put sand and marbles in a glass tube, then ran sewage over the lot. At first, he noted the sewage came out unchanged. But after a time, the effluent became clear. When he applied chloroform to the sewage, the effluent ran muddy again. Only after all traces of chloroform were eliminated did the effluent again emerge clear. From this experiment he concluded that living organisms in sewage were responsible for purifying. In England, Frankland ran sewage through different types of soil, and he found that both loam and sand had a purifying effect. He believed air contributed to the purifying action.

From its inception the scientists at the Lawrence Experiment Station concentrated on developing different sorts of filtration techniques. The first thing Mills did was to build ten big tubs, filled with different kinds of New England soil, then poured sewage over them. He buried a dead dog in one of the tubs to see whether its decaying flesh might not help in the purifying process.

Although it was not clearly understood at the time, bacteria contained in sewage fed on the sewage. Thus, the sewage was transformed into inert, non-organic substances. The idea of treating sewage consequently is to find ways of speeding up this natural process. When sewage was passed through sand and gravel, it was discovered that it changed form rapidly. The bacteria clung to the particles of stone and sand and fed on sewage as it passed through, helping to purify it. The next step was to pass sewage over rocks or rock-like forms, and to hasten the decomposition by blowing air through it all. (Pasteur had discovered that certain bacteria thrive on air while others do not.) These many experiments led to the development, by the turn of the century, of the trickling filter. In 1970, the trickling filter is still considered among the most effective forms of "secondary" treatment, a kind of sewage treatment most towns and cities in America still do not have.

While the trickling filter is regarded as an excellent way of hastening the natural purifying process for sewage, big cities tend to reject the method because of the great amounts of space required. Instead, the most widely accepted form of advanced treatment in 1970 is another method, also developed at Lawrence about 1910, and called activated sludge. In this method, sewage is run through big tanks. The sludge which settles to the bottom contains billions of bacteria. Some of this sludge is blown through the liquid sewage as it comes into the tanks, and the bacteria feed on it. An activated sludge plant may only take up a few city blocks, where trickling filters would require large acreage.

In the 1890s, Lawrence experienced a terrible bout of typhoid, and Mills, who was convinced the disease was carried in the drinking water, decided to try out some of the filtering devices they had developed for treating

sewage in filtering the drinking-water supplies. His experiments proved successful, and filtration of the drinking water at Lawrence coincided with a sharp decline in
both the rates of typhoid and the deaths due to the
disease. The Lawrence researchers also experimented
with chlorine after it was developed in Germany.
Chlorine was introduced into the United States about
1912. In discovering that drinking water could be
purified by different filters, and made doubly safe
through chlorination, interest in pollution declined.

Following the experience of the English, sanitary-
engineering theory gradually changed course and
adapted dilution as a principle on which to base new
sewage systems. The dilution theory was first put into
widescale use at Chicago with the construction of the
Chicago Drainage Canal. The sewage from Chicago was
pumped directly into Lake Michigan, from which the
city took its drinking-water supplies. The result was
that people would find pieces of garbage and filth in
their drinking water. In the 1870s, there was discussion
about purifying the Chicago sewage by filtering it
through sandy areas to the south of the lake, but that
idea was rejected because of the expense. Instead, in
1889, the state of Illinois authorized construction of a
drainage canal, or open sewer, which carried sewage
from Chicago to the Illinois River, and that in turn
emptied into the Mississippi. Rudolph Hering, a well-
known engineer, designed the canal. Hering had
traveled extensively in Europe, and like other members
of the new profession of sanitary engineering he was

conversant with practices in England, France and Germany. The irony of it was that Hering does not appear to have had much faith in dilution. He warned repeatedly about the dangers of flushing sewage into bodies of water.

The building of the Chicago Drainage Canal marked the beginning of the process which turned the Mississippi River into an open sewer. Many large cities on the eastern seacoast and in the Middle West adopted the dilution theory as a result of deliberate planning, or they continued to dump filth into rivers because of custom. In 1906, the New York State legislature created a Metropolitan Sewerage Commission, which over a four-year period completed a study of the five boroughs of New York, reaching basically the same conclusions as did the Federal Government in its survey of New York waters during the 1960s. In 1910 the Commission reported: "Practically all waters within 15 miles of Manhattan Island are decidedly polluted." The pollution was greatest between the Narrows, Throgs Neck, Mt. St. Vincent and the mouth of the Raritan River. "Gowanus Canal and Newtown Creek and the Passaic River are polluted beyond the limits of toleration. The Harlem River, particularly at its southern end is, at times, little else than an open sewer." Then, as now, it was obvious that it made little sense to dispose of sewage by flushing it into a tidal estuary such as the Hudson or East Rivers, where the waters churn back and forth every six hours. The Commission made elaborate float studies of New York Harbor; it concentrated work on the East River.

The report says: "Most of the water flows back and forth like the pendulum of a clock without escaping to the ocean or to Long Island Sound. A buoy which was made to float, except for a small tell tale, just below the surface of the water, was followed back and forth for three and a half days in the East River. At the end of this time, it had traveled 108 miles without passing out of the stream. It returned several times to the locality where it had been set adrift.

The great bulk of sewage from Manhattan Island, the four other boroughs and suburban New Jersey communities in 1910—as now—was poured into New York Harbor. Manhattan, then as now, ran its sewage from dwellings through pipes into the rivers, with no attempt to treat it. Six-hundred-million gallons were put into rivers each day, as compared with 1.3 billion gallons a day now. There was an awareness then of an increasing sewage problem, and the solution proposed by the neighboring communities, such as the Passaic River Sewerage Commission, was to extend their outfall pipes farther into New York Harbor. The Metropolitan Sewerage Commission warned of the danger, should this practice be followed. Nobody paid any heed.

At that time, metropolitan New York comprised forty different cities with 2,903 miles of sewer pipes. Most of it was combined sewer pipe, carrying both sanitary and storm run-offs. There were several purification plants, although the Commission found all but one of them in rickety and next to useless condition.

The Metropolitan Sewerage Commission concluded that pollution was a serious problem: "The placing of floating bathing establishments near sewers, as well as the washing of the shores of the bathing beaches with sewage is a decided nuisance and menace to the health of the patrons thereof. The present popular conception of a nuisance is undoubtedly more strict than it has been in the past, and it seems reasonable to suppose that future sanitary requirements will establish higher standards than now prevail."

3 | Pollution Control

In the United States, during the first decade of the century, Progressives and Republican insurgents calling themselves "conservationists" joined forces in a concerted campaign to develop natural resources in what they believed to be an efficient manner. President Theodore Roosevelt, who led the campaign, termed conservation, "the chief material question that confronts us, second only—and second always—to the great fundamental question of morality." Roosevelt's interests were often utilitarian, and reminiscent of Chadwick's. To the President and his followers conservation meant devising a strong central nationalist system for controlling development of natural resources. The conservationists were concerned with proper management practices, and to achieve those ends often sided with large corporate interests.

Gifford Pinchot, the chief forester in Roosevelt's Administration, allied himself with the large cattle corporations against settlers and homesteaders in his determination to lease Federal range for grazing. Roosevelt complained that "people refuse to face squarely the proposition that much of these lands ought to be leased and fenced as pastures and that they cannot possibly be taken up with profit as small homesteads."

Roosevelt first opposed private utility operations on Federal property, and then worked with the power companies to help them develop sites. His Interior Secretary, James R. Garfield, a leading conservationist, aided oil prospecters in withdrawing lands from agricultural uses and reserving them for petroleum exploration. Led by Pinchot, the Roosevelt Administration backed groups of commercial and manufacturing interests who were eager to build deep inland waterways which businessmen hoped would reduce the high cost of shipping by rail. Small lumber men were bitter about Roosevelt's sell-out to the big lumber companies in his forest policy.

In his book, *Conservation and the Gospel of Efficiency,* Samuel P. Hays writes of the conservation crusade. "The President, Pinchot and Garfield carried their interest in efficiency into a variety of fields other than natural resources. They emphasized, for example, the value of large-scale business organization, and warned that anti-trust action might impair increased production. They had no quarrel with bigness as such. Instead in a manner similar to Thorstein Veblen, they distinguished between the efficient production engineers and the adventurous profit-taker who sacrificed a wider distribution of cheaper goods for speculative investment returns. Such profits, declared Pinchot, were a 'tool levied on the cost of living through special privilege.' Speculative profit-taking produced as much waste as did competitive exploitation of natural resources. In either case conservationists strove to encourage the greatest possible production of material goods at the lowest cost. Just as planning must replace competition so that manufacturers could produce less waste, so regulation must prevent financial freebooters from destroying industrial efficiency."

While many people remember the conservationists as forces of good, fighting it out with the special interests to save the land from savage rape, it must be remembered that they were not socialists. They objected to bad business, not to the idea of business. In practice, large corporations could most easily afford to take up the planning and efficiency concepts advocated by the conservationists.

That coalition between big business and big government, formally established during Theodore Roosevelt's Administration, has remained in force ever since. By the late 1930s, the conservationist campaign settled on water pollution.

During the New Deal, pollution became a serious problem, but there were deep differences of opinion over what to do about it. Middle Western Republicans, led by Karl Mundt, then a congressman from South Dakota, now a senator, and conservation groups headed by the Izaak Walton League, argued for Federal water-quality standards and a stiff enforcement program. The Republican conservationists viewed pollution as part of water-resource development, and they believed it should be handled by the United States Army Corps of Engineers, which was developing waterways. But the Franklin D. Roosevelt Administration opposed the scheme for Federal standards, and instead suggested a plan for handling pollution on a regional basis through river-basin commissions or authorities which would manage development of natural resources resources along the rivers and their tributaries. The administration was enthusiastic about the Tennessee Valley Authority and hoped it could serve as a model for other regional ventures.* In addition, by using Federal funds the Roosevelt government was helping industry recover from the Depression, and was not anxious to press business into cleaning up its wastes.

Water pollution was most severe in the heavily industrialized eastern section of the country, a rectangular

---

* TVA is now the biggest supplier of low-cost power in the country. Because of the pressure to produce such large quantities of cheap power, TVA uses the least expensive fuel available: strip-mined coal, gouged out of the hills of eastern Appalachia. Ironically the Roosevelt conservation model is now the biggest cause of pollution in Appalachia.

strip bounded on the west by the Mississippi River with
Chicago at the top and St. Louis on the bottom, and in
the east along the coast from Boston to Washington.
New Deal expenditures for sewers under the Public
Works and Works Relief projects rose from $47 million
in 1932 to $257 million by 1938. By then over half the
urban population had sewer systems. At the turn of the
century the figure was 3 percent.

A National Resources Committee, led by Harold
Ickes, Secretary of the Interior, made several surveys of
water pollution, and in 1938 concluded: "During the
past six years the proportion served by sewers has increased more rapidly than the total population or urban
population. If present inducements to sewer construction are continued, it is reasonable to assume that the
recent trends will continue and that in a few years all
of the accessible population will have been reached
by sewers." The Ickes group, however, acknowledge
that these predictions did not include industrial wastes,
some of which were not susceptible to any sort of treatment. Treatment for other industrial wastes was viewed
as too expensive for serious consideration. The economic
benefits of the industries in terms of jobs they created
outweighed their menace as polluters.

The National Resources Committee opposed more
Federal pollution legislation, and instead urged a program of grants-in-aid to localities to help them build
sewers, and loans to industry so they could handle wastes.
(The total cost for a thorough clean-up in each category
was estimated at $1 billion.) The committee came down

hard against the idea of national water-quality standards with a simple endorsement of the dilution principle: "Streams are nature's sewers." The report said: "The key problem in planning for pollution abatement is to find the standards in each section of a stream which express the best balance between the stream's use for receiving and assimilating sewage and other waste and its use for other purposes, aesthetic and economic."

Instead of standards, the Ickes committee encouraged regional commissions along the lines of the Interstate Sanitation Commission, a compact between New York and New Jersey, approved by Congress in 1935 to establish standards for coastal waters in New York and New Jersey. While the National Resources Committee urged states to tighten their laws, it warned: "Enforcement can proceed no faster than basic research reveals the effective means of treatment, or than financial arrangements can be completed to construct the necessary works."

Conservationists assailed the Ickes report as a publicity stunt for industry. The committee had acknowledged its debt to several trade groups, including the Manufacturing Chemists Association, American Petroleum Institute and the Iron and Steel Institute. Nonetheless, Roosevelt accepted the Ickes line and stuck to it.

During the thirties the main congressional opponent of a strong Federal pollution program was Alben Barkley, the Democratic senator from Kentucky who became Truman's vice president. Barkley was a defender of the coal operators of the Ohio River Valley, a main target of the conservationists and one of the most badly polluted areas of the country. In 1936, Augustine Lonergan, a Democratic senator from Connecticut, sponsored legislation which would have created Federal water-quality standards and given the government injunctive powers to halt pollution. Barkley in the Senate and Fred Vinson, his Democratic colleague from Kentucky in the House, sought to sidetrack Lonergan's bill with their own proposal to create a division of water pollution in the Public Health Service, which could make investigations and give out sewer grants to municipalities. There were no provisions for setting standards or for Federal enforcement. The Barkley-Vinson pollution bill produced a torrent of ridicule. Charles Faddis, a Democratic representative from Pennsylvania, declared, "This is not an anti-pollution bill. . . . This is a bill to once more lull the forces that are working against pollution into sleep to get their minds off the subject of the real cause of pollution . . . pollution from industrial waste. There is not a member of this House but knows what is causing stream pollution. What we need is legislation with teeth in it to prevent this pollution. This is a bill for the promotion of bureaucracy."

Lonergan was joined in the fight by Joseph Bennett (Champ) Clark, the Missouri congressman, but Lonergan's proposals were killed in conference, and a meaningless bill, devoid of standards or injunctive powers, was finally approved. Even so, Roosevelt vetoed the final act, claiming it denied him adequate control of the program.

In January 1940, Karl Mundt sponsored a bill, the Navigable Waters Anti Pollution Act, to be administered by the Army Engineers which would give the Federal government considerable power in cleaning up pollution. The engineers would take charge of classifying navigable waters into sanitary water districts, fixing standards for purity in each district, and establishing and promulgating minimum requirements for treating polluting material before it was discharged. Where industry compliance with the standards was impossible within two years of the bill being passed into law, the engineers could grant the industry an extension for up to five years. However, compliance would be considered impossible only if no adequate method of disposal was known to the polluter. The burden of proof would be on the industry which sought an extension. Disposal of waste into United States navigable waters would be a violation of law, and at the request of a local sanitary district or the Secretary of War, the government would be required to take legal action.

Despite the fact that Mundt's bill would have placed authority for controlling pollution in the hands of the engineers, whose water-development schemes often caused pollution, it represented a real effort to attack pollution within a development policy. The Corps has always regarded pollution abatement as a secondary matter which must not be permitted to interfere with the primary job of building dams, dredging channels and opening harbors.

Mundt's bill was killed in committee, and again the leading antagonist was Barkley. In March 1940, Barkley introduced the same bill he had put in against Lonergan. It passed the Senate, but the House tacked on Mundt's legislation as an amendment. The two different bills went to conference, where the issue died.

The conservationists were always bitter toward Roosevelt. In 1940, the Izaak Walton League's president wrote a government official: "It has always seemed to the Izaak Walton League and numerous other conservation organizations, more than passing strange that this liberal administration concerned with the best interests of the great mass of our people should have allowed itself to be jockeyed into the position where they are actually supporting pollution of our waters." But Roosevelt was not deterred. His position was clearly stated in a message to Congress in 1939. He said then, "Federal participation in pollution abatement should take the general form of establishing a central technical agency to promote and coordinate education, research, and enforcement. . . . It should be supplemented by a system of grants in aid and loans."

After the Second World War, in 1948, Barkley was back with the same old bill. This time he sponsored it along with Robert A. Taft: it passed and became the first Federal pollution law, called the Water Pollution Control Act. The Act declared pollution was best handled at the local level, but it required the US Public Health Service to provide technical information to the states and to help them coordinate research activities. It also made available token sums for building sewers, and if requested by states, the Federal government could take a hand in enforcement. The actual appropriations declined over the years, from $3 million in 1950 to less than $1 million in 1955. The Act was to expire in 1953, but was continued until 1956. While the legislation did not provide much in the way of money, enforcement or policy, it did finally—after more than a decade—establish the principle of Federal activity in the area of pollution.

In 1956 John Blatnik, a young liberal congressman from Duluth, Minnesota, took an interest in the water-pollution program and, together with a handful of men, he laid out a stronger pollution law, and has played a major role in shaping and directing the pollution programs ever since.

Duluth, at the western tip of Lake Superior, is at the center of the depressed iron mining ranges. The economic health of the region is closely tied to water, in part because it is beautiful country which attracts vacationers, but more important because of the ships and barges which steam in and out, connecting the inland mines and manufacturers to the rest of the country and to the world. As a freshman congressman Blatnik sought to develop water resources. He vigorously pushed research into the possibilities of making iron from low-grade taconite ore left in the Mesabi Range and previously considered unusable. The research paid off and refining taconite increased employment and commerce.

In the early 1950s, Blatnik was appointed to the

Public Works Committee of the House, and within that committee, was made a member of the subcommittee on Rivers and Harbors—one of the most powerful subcommittes in the Congress, deciding on dam and irrigation projects, harbor dredgings, canals and river projects. Rivers and Harbors works closely with the Army Corps of Engineers. When Blatnik joined the House, membership was important to him because he could take a direct hand in lobbying for the creation of the St. Lawrence Seaway. As a member of the subcommittee, Blatnik traveled down the Mississippi, and around the Gulf and East Coast ports. He was struck by the filthy harbors of the lower Mississippi. In 1955, Blatnik became chairman of Rivers and Harbors, and his first business was the Barkley-Taft Water Pollution Control Act legislation, which was due to expire.

Blatnik gave the task of handling the pollution legislation to Jerome Sonosky, a young lawyer on his staff. Sonosky, in turn, worked out the details with Dr. Gordon McCallum, head of the pollution division within the Public Health Service, and McCallum's new assistant Murray Stein. Sonosky also consulted with Hugh Mields, lobbiest for the American Municipal Association (the organization has since changed its name to the National League of Cities). Together these men developed the legislation and kept in close contact with Blatnik's congressional lieutenants—Robert Jones of Alabama and James Smith of Mississippi. The final bill called for grants to municipalities to help them build sewers, and a new enforcement scheme which would enable the

Federal government to pressure polluters into cleaning the water. The legislation frightened Southerners who took it as another move by the Federal government to encroach on states rights. Officials in charge of local pollution programs resisted the idea of Federal enforcement since they feared encroachment of other bureaucrats in their work, and it antagonized fiscal conservatives both within Congress and in the Eisenhower Administration, who were anxious to cut back Federal spending, not increase it.

The major political obstacle to the legislation was Robert Kerr, the Oklahoma senator who was head of the Rivers and Harbors subcommittee in the Senate Public Works committee. Kerr promoted himself as a staunch supporter of water-pollution programs, but behind the scenes he worked to gut the Blatnik bill in the interest of the oil industry, where he had heavy financial ties. Kerr pared down the amount of money for sewage grants from the $100 million approved by the House to $50 million.

In its final form, the Federal Water Pollution Control Act of 1956 authorized $500 million over ten years in matching grants to help localities construct sewage disposal plants. There was a limit of $50 million a year. Each proposed project could only qualify for 30 percent of its total cost in government funds, and as a sop to rural and small-town House members, 50 percent of the money was to go to municipalities of 125,000 people or less. Under the legislation, if the Surgeon General of the US Public Health Service "had reason to believe" pollution of interstate waters was occurring, he could call a conference of officials. The conference was meant to bring Federal, state and local officials together in working out an abatement scheme. If no action occurred within six months of the conference, the secretary on request from the offending or offended state could request the Attorney General to bring suit against the offender. Finally, the act authorized increased research, grants to help states build up pollution-control agencies, and established a Water Pollution Control Advisory Board.

President Eisenhower was against the pollution law because of its spending provisions, and the Budget Bureau froze the funds for several months. Southerners attempted to gut the grants provisions in the appropriations committees, but they were beaten off on narrow votes. In 1958, Eisenhower asked Congress to discontinue sewer grants. He believed states should finance sewers, and in order to help them finance public-works projects of this and other sorts, he proposed abolishing the Federal telephone tax and instead allowing different states to collect a telephone tax. Blatnik's reaction to these proposals was to put in a bill which increased sewer grants from $50 million to $100 million a year. A bill increasing spending to $90 million a year eventually passed both houses, whereupon Eisenhower declared: "Because water pollution is a uniquely local blight, primary responsibility for solving the problem lies not with the federal government but rather must be assumed and exercised, as it has been, by state and local governments. This being so, the defects of HR 3610 are apparent. By holding forth the promise of a large scale program of long-term federal support, it would tempt municipalities to delay essential water pollution abatement efforts while they waited for federal funds." The House was not able to over-ride the veto.

The change-over from Republican to Democratic Administrations in 1960 meant little difference from the standpoint of environmental pollution. Under President John F. Kennedy the White House took little interest in Water Pollution. Abraham Ribicoff, Kennedy's first secretary at HEW, vigorously opposed giving the tiny pollution division more authority. There were further efforts by Blatnik to increase the amounts of money for sewer grants, and to provide more importance to water pollution, but these efforts came to nothing.

The 1956 law remains as the basic Federal pollution statute, and despite the amendments added since, it is a relatively weak law—far weaker than those adapted in England a century ago, and weaker than the proposals made by Lonergan and Mundt in the 1930s. The burden of proving pollution exists is on the Federal government. But even if it proves pollution exists, it cannot stop it.

On one level the act represents a stand-off between the states and the Federal government over control of water resources. On quite another level, it precipitated battles for control of pollution programs among warring sects of bureaucrats. While the quarrels are cast in a setting different from Chadwick's times, they nonetheless are in many respects similar to his long, bitter battle to extend the apparatus of the central government to local communities.

In many states water-pollution programs were organized and administered through the public-health services, and while sanitary or civil engineers work within these services, they are controlled by medical doctors. The physicians' primary interest in pollution has always been in breaking the disease chain between drinking water and sewage. The development of sand filters and application of chlorine to the water supplies made it possible to ensure a clean drinking-water supply. As the rate of water-borne disease—typhoid, diarrhea, etc.—declined, the public-health doctors understandably lost some of their enthusiasm for pollution control. At the same time, with the Federal programs slowly expanding, the public-health bureaucracy was not anxious to let go control of pollution programs, which increased the size of their budgets and their importance.

Where a health problem went beyond state borders, the state-medical officers consulted their colleagues elsewhere, and they came to depend on the Federal public-health service for information, coordination and consultation in instances where emergency was widespread. But always these relationships were "consultation" arrangements. The idea which emerged from the Blatnik bill, that Federal public-health service officers were charged to begin enforcement procedures, possibly even court actions, against state health departments for failure to enforce the pollution control laws, was considered an affront by some and madness by others. It smacked of socialism; it was unnecessary since water-borne disease rates were low and under control, and it was insulting—pitting members of the profession against one another.

Thus, the public-health service doctors, as well as sanitary engineers working within these agencies, found themselves the target of the Federal attack. In defending themselves they became the spokesmen for vested interests in pollution—heavy industry, inept sanitation departments, the Corps of Engineers, engineering societies and so on.

As the government began to apply the enforcement provisions of the act, calling conferences of state and local officials and industrialists in an effort to win agreement on abatement plans, it discovered the real antagonists were the state and city agencies, which were meant to control pollution but in reality were pursuing policies which caused pollution. Federal officials charged with enforcing pollution laws took the position that their professional bonds to the state and local health doctors were more important than the laws they were charged with administering. And thus, within the Federal bureaucracy the program broke down in quarreling among professional groups, especially among the doctors of the public-health service and the lawyers within the enforcement section of the pollution division.

One of the most frustrating examples of professional bickering occurred in New Jersey on the Raritan Bay.

Travelers going in and out of New York City from New Jersey pass through this industrial mire, an area of stinking bogs, with flames and putrid smoke belching into the sky from chimneys of oil and chemical companies. At the feet of these industrial engines, running amidst a clutter of towns and small cities, flows a foul-smelling sewer, the Arthur Kill, into which industries steadily pour all manner of refuse. The Kill lies between the New Jersey shore and Staten Island, terminating on the north in Newark Bay. At its southern outlet the accumulated filth is fed into the Raritan Bay, a body of water thirty square miles in area located beneath the sprawl of New York Harbor and designated since 1937 by New York and New Jersey as a playground for the 15 million people living in and around the city. Beneath this huge cesspool, close to where the Arthur Kill spills out its contents, other sewer pipes converge beneath the water's surface spewing forth 50 million gallons of human and industrial waste each day. In all, the sewage from 1.2 million people is pumped into this bay every twenty-four hours. Large amounts of inorganic waste also are deposited. This mass of putrefaction oozes about the New Jersey and Staten Island shores for several days, washing the beaches with great quantities of fecal bacteria, closing out the light and consuming the oxygen required by fish and other forms of marine and animal life, before sluggishly moving seaward on the outgoing tide.

Raritan Bay was designated Class "A" waters by the
Interstate Sanitation Commission, the organization
formed in 1935 by the states of New York and New
Jersey to regulate interstate water pollution. The Roosevelt Administration regarded it as a model for future
river-basin compacts. Class "A" means that the waters
are to be used primarily for bathing, boating, fishing
and other recreational purposes, as well as for the maintenance of a shell-fish industry. (Pollution has forced
the closing of all but a small portion of the once pros-
perous shell-fish beds on the Raritan.) State health
commissioners are members of the Interstate Sanitation
Commission working independently and through the
commission to enforce water standards.

The numerous tributaries, twisting tidal currents and
the great increase of population and multiplicity of industry have combined to make abatement of pollution
on the Raritan difficult. Even so, efforts by the states
and the commission have been so belated and so feeble
that the Bay is one of the worst sewers in the nation. In
1961, an outbreak of infectious hepatitis was traced
to clams taken from the Raritan Bay, bacteria standards
for which had not been enforced by the states. The
hepatitis outbreak became an especially sore point because specialists in the US Public Health Service had
carefully traced the hepatitis outbreak, made what they
believed to be accurate correlations between victims and
the clams, and offered what they considered to be over-
whelming scientific evidence to prove the point. Nonetheless, the senior US health officials in New York, who

were medical doctors, simply refused to send warning memoranda to their colleagues in the New Jersey state-health department, lest they be embarassed.

Acting under authority granted it by the Water Pollution Control Act of 1956, the Federal government, against strong opposition by the states, attempted to improve conditions in Raritan Bay.

A Federal laboratory was thrown up at Metuchen, New Jersey, and a team of specialists put aboard an old United States tug. Quietly, with no interest by press or public, the government's twelve-man office began its war against the New Jersey state bureaucracy, and against the public-health service.

The first enforcement conference on the Raritan was held in August 1961. At this session, state health officials as well as the Interstate Sanitation Commission, through which they sometimes functioned, protested that their pollution-abatement programs were effective, that the seventy-three cases of hepatitis "hardly presents the 'national' interest that would justify enforcement action by the Public Health Service," and finally suggested that one of the major purposes of the waters in question was for the disposal of waste. William C. Cope, then chairman of the Interstate Sanitation Commission, declared, ". . . Surveys by state and city agencies indicate that they (Raritan Bay waters) are now, and for some time past have been, in very good condition for virtually every use that can be made of them, including those requiring a high degree of purity such as fishing and bathing. . . ."

As far back as the Metropolitan Sewerage Commission, there had been reports that the Raritan Bay was polluted. In 1961 the water off Perth Amboy, a New Jersey public bathing beach, was particularly warm because 200 feet off the beach front, rising in a boil, spouted the city's sewage outflow pipe, bathing swimmers and the beach with fresh human feces. A few miles away, the town of Tottenville, on Staten Island, dumped raw sewage from 4,000 people into the bay waters and several miles east up the Staten Island coast, the public bathing beaches were washed in the sewage flows coming through the Narrows from New York Harbor. At one of these, South Beach, the City of New York enacted a little drama every year. The City Health Department warned the people not to swim there because the water was polluted. Then the city parks department came along and opened the beach staffing it with lifeguards.

There were indications that policies followed by both the Interstate Sanitation Commission and the state health departments tended to increase the complexity of the pollution problem. Since 1958, for example, the Middlesex County Sewerage Authority, one of the largest contributors of waste to Raritan Bay, has been permitted to pump its effluent into seven feet of water in a dead spot in the Bay, adjacent to heavily populated areas and numerous bathing beaches. As a result, the sewage lies in this dead spot for several days before moving out to sea. To make up for the shallow water, the Authority dug a hole thirty-five feet deep in the sand beneath the end of the pipe. Like many other sewage plants in the Raritan Bay and elsewhere, the Authority provided so-called "primary" treatment. The sewage was collected, in a big tank, permitted to settle with the solids sinking to the bottom, and the liquids then poured into the Bay. To kill harmful bacteria in the effluent, it is chlorinated before being expelled into the water. The government claimed the chlorine was not administered in sufficient quantities to kill the bacteria, and great volumes of effluvia reduced the oxygen content of the water below the commission's own standards.

The inabilities of the states to compel industry to clean up its wastes are well known. But the states of New York and New Jersey seem to have run up against particularly bad luck in attempting to persuade hundreds of industries along the Arthur Kill to install treatment plants for their inorganic and organic wastes. Since 1955, the Interstate Sanitation Commission has made certain studies of the Kill. Each one showed definite progress was in sight. The industries, never named, were being consulted.

Of course, this was a delaying tactic. Pollution could have been pinpointed by sampling the outflow pipes of the different companies, and then requiring them to install proper equipment. Moreover, Chairman Cope and his successors consistently maintained that the Arthur Kill really had no effect on the Raritan Bay at all, but rather, expelled its heavy pollution load to the north into Newark Bay. However, this view was contradicted by the commission's own research, which reported that dye, floated at any point in the water to show current direction, rapidly began to leave the Kill through both northern and southern ends. A Public Health Service study of the Kill showed, that under the most adverse conditions, anything placed at the Raritan Bay end of the Kill would float all the way up on one tide and come back down in short order.

When the Public Health Service published the results of this study, the Interstate Sanitation Commission fell silent on its claims of current directions of the Arthur Kill. Orders were issued through the state health departments, directing several municipalities and industries, most of them on the New Jersey side of the Kill, to improve markedly their sewage-treatment facilities within one year's time. In announcing these orders as proof of its effective and progressive plans for water-pollution control at the second session of the conference in May 1963, the Commission and the states pleaded with the Federal government to quit the Bay. The government studies, they claimed, were a scant contribution to the knowledge of the areas. But the government declined to take its tugboat away, claiming in turn that its studies were not then complete.

Shortly after the second session of the conference was concluded, New Jersey announced that one year would be too much of a strain on the offending industries, and that at least two years of planning time would be necessary before a concrete timetable could be developed. There are three large industries which contribute to the bulk of the pollution along the Arthur Kill, Humble Oil, American Cyanamid and General Aniline and Film.

In seeking to persuade the states to clean up the Raritan Bay, the government lawyers were not only faced with layers of state bureaucrats, who acted as if they were paid public-relations hands for the industries, but behind them they ran up against their own employer, the US Public Health Service. The PHS provided public funds to support the New Jersey and New York public-health departments and gave money to the Interstate Sanitation Commission for the bogus research it was carrying out.

Murray Stein, in charge of enforcing water-pollution laws, was sandwiched between bureaus of the US Public Health Service. Stein could make forays against pollution so long as they never reached beyond the conference level. Even before the second session of the Raritan Bay conference began, Stein was trying to strike up some interest in the subject among reporters. The US Public Health Service shifted the site of the conference from the Federal courthouse at Foley Square to the offices of the Carnegie Endowment because they thought Foley Square might seem unduly oppressive from the state point of view. Then the Public Health Service doctors sought to stop the issuance of an innocuous press release announcing the conference, because of what the regional office believed to be "sensitive" Federal-state relations. Thomas R. Glenn, Jr., executive secretary of the Interstate Sanitation Commission, believed the commission was best able to make progress when news of water pollution "ends up in the back of the newspaper. The Commission has a reputation for being reasonable," he added.

In 1965, a third session of the Raritan Bay Conference was held, and more plans and reports were issued. The New Jersey communities were by then equipped with pilot plants of sewage-disposal schemes. By 1969, the Federal government was contemplating holding another conference about the Raritan Bay, and New Jersey, under a new leader, Richard Sullivan, had put the industries and municipalities under orders to clean up. Time-tables, when developed, ran into the 1970s.

It is nearly ten years since the Federal government entered the Raritan Bay to arrange for a clean-up; nothing much has happened except that population and industry have increased, adding to the levels of pollution.

By early 1962 Stein was thoroughly frustrated. He could not move the state health doctors, and he believed his work was being sidetracked by other doctors within the US Public Health Service in Washington, who did not care for his aggressive attack on pollution. Although it was legally possible to move the pollution office out of the Public Health Service, neither Ribicoff nor his successor Anthony Celebrezze would do so.

For his part Hugh Mields at the National League of Cities was anxious to persuade the Federal government to make available substantial amounts of money for building sewers in cities, then as always pressed for money. Mields and Stein together visited Congress, looking for congressmen and senators who could be interested in supporting pollution legislation. And eventually they met with positive response from Senator Edmund Muskie of Maine.

As Governor of Maine, reflecting the same sorts of interests as John Blatnik, Muskie had worked to improve the stream classification system within the state, and had fought the Eisenhower funding scheme. Because of increasing pollution in its waters two of Maine's major industries, fishing and shell-fishing, were in rapid decline.

Stein and Mields met with Donald Nicoll, Muskie's administrative assistant. The three of them informally wrote amendments to the pollution act. In an immediate sense the amendments reflected the interests of this small group of individuals. Under the proposed legislation the water-pollution division within HEW would be sprung away from the Public Health Service and established as a separate administration beneath the secretary. Sewer grants were to be increased to $500 million a year and a special program begun to finance the separation of storm and sanitary sewers. All three agreed on a section which would require establishment of Federal water-quality standards.

In 1963, both Kerr, chairman of Rivers and Harbors, and Dennis Chavez, who chaired the Public Works committee, died and Pat McNamara, a Michigan Democrat, ascended to the chairmanship. McNamara hired Ron Linton, a former Michigan newspaperman, and

Kennedy organizer, as chief clerk of the committee, and abolished Rivers and Harbors as a separate subcommittee in order that he might keep a tighter rein on the pork-barrel legislation himself. Partly to placate Muskie, who had been interested in becoming chairman of Rivers and Harbors McNamara created a special subcommittee on air and water pollution and made Muskie chairman. That subcommittee did not have a staff, and instead Linton and Nicoll ran it. It was never a partisan affair since the ranking Republican member was Caleb Boggs, of Delaware. Muskie knew and liked Boggs from the time they were both governors, and Boggs's assistant, William Hildenbrand, became a third member of the de facto subcommittee staff. That group also included Mields, and of course, Stein, who throughout this entire period was the single most vigorous proponent of pollution reform.

This tiny group of congressional and administrative bureaucrats created a virtual political organization which legislated and administered water-pollution legislation throughout the 1960s. This was a game Democratic politicians could risk playing since they were in effect connecting the Federal government, under Kennedy and Johnson, to its political base in the Democratic-controlled city machines, shifting more of the pork from the countryside to the city. The key standard setting provisions of the act represented a sort of compromise solution among competing interests—between conservationists who wanted uniform policies administered from a central government dedicated to rational resource development, and the older New Deal Democrats who were anxious to experiment with regional governments. Under the standard setting mechanism, as described in the Act the government publishes criteria for water-quality standards. The states then classify the water, and set standards which are submitted to the Federal government for approval. If they are not approved, then the dispute goes to a hearing.

Muskie's pollution amendments met with little resistance in the Senate where they first were passed in 1963 by a wide margin. However, the bill was locked up in Blatnik's Rivers and Harbors subcommittee for nearly two years because Blatnik could not budge representatives from Louisiana and Florida who were responsive to oil interests. The bill, however, was reported out of committee and became law in 1965. Thus by 1965, pollution legislation was back on the course of establishing a strong central control, about where Mundt's bill would have put it in 1940.

But even after these major reforms—made in the public interest—with promises all around, water pollution amounted to nothing more than a private preserve for opportunistic politicians. During the mid-1960s Stewart Udall, the Secretary of the Interior, set out to capture the water pollution program from HEW. Udall wanted to increase his political prestige within the government by administering a pollution program based on a grants-system, then contemplated at $1 billion a year, which would give him leverage in the Eastern cities as well as in the mountain and Western states where Interior traditionally has its authority. The Interior Department usually is regarded as a regional department, a source of pork for different Western interests, most of them concerned with oil and gas, mining, ranching or timber.

Since Udall could not control the different Interior bureaus representing the different vested interests in the extractive industries, there was no reason to believe he could do any better with water pollution. Moreover, Johnson himself was handling oil policy from the White House through his aides DeVier Pierson and Marvin Watson. Putting pollution programs into the Interior Department, which was controlled by oil, gas, mining, ranching, etc., the very interests which caused so much of the pollution, would in all likelihood serve industry, in particular oil, very well indeed. For whatever reason, Johnson approved Udall's take-over scheme, and the Secretary of the Interior went to the Capitol to argue for the plan with Muskie, Blatnik, and with Abraham Ribicoff, who chaired the government-operations committee. Government operations needed to approve the final change-over.

Udall knew everyone involved, either from his days as a congressman or as a member of the cabinet. He argued that while Interior might not at first appear to be a proper place for water pollution, he would give it his full personal attention, promising to spend one third of his time on the subject. Neither Blatnik nor Muskie much cared for the idea, but neither did they have much inclination to fight the President, if he really wanted to make the change. At that point, in 1965, there still was no real constituency for water pollution outside the interests represented by the handful of men who wrote and promoted the legislation.

More to the point, Udall worked a deal. If Muskie and Blatnik agreed to the move, committee assignments would remain as they were. Interior matters are usually handled by the Interior Committees. Now two public-works subcommittees, with quite different sets of interests and constituencies from the Interior Committee could influence half the budget of the Interior Department. At is turned out, Muskie emerged with a better deal than Blatnik. After the move, Udall's staff sat down with Muskie's staff and the two groups interpreted the laws which Muskie's people had written. Thus, Muskie's interests really became intertwined with Udall's and the legislator and his office became the administrator as well.

Once Udall gained control of the pollution program, he lost interest in it, and its administration drifted aimlessly. The move out of the Public Health Service to Interior meant that many of the Public Health Service technicians dropped out. They refused to go to Interior for fear of losing their perquisites in the PHS—high retirement income, etc. Udall took the pollution section and combined it with another unit involved in water research. The whole business was put under control of Frank DiLuzio, a research man from the Atomic Energy Commission. In the shift-over, Stein was passed over for promotion on the grounds he did not possess enough administrative abilities and all around experience for the job of commissioner. (Stein holds virtually the same job in 1970 that he did in 1956.) Instead the commissioner's job was given to James Quigley, a former assistant secretary at HEW and an old friend of Udall's.

Almost immediately a squabble broke out between Quigley and DiLuzio over the standards. It came to DiLuzio's attention that some states were approving water-quality standards which would permit actual lowering of quality of some untouched streams. The standards had become, in effect, a way to license pollution. DiLuzio pursued the idea of instituting a non-degradation standard, through which the states would be required to guarantee that they would not lower the quality of water in any stream below existing levels. DiLuzio argued for the non-degradation principle, while Quigley was against it. He thought it was being too harsh on industry.

Finally Udall and the quarreling pollution officials worked out an arrangement where it was made to appear that the Secretary approved the idea of "non-degradation," where in fact he really wasn't asking the states to do much of anything. Udall sought to persuade the states to adopt a so-called non-degradation clause which says they are pledged not to lower existing water quality. That resulted in a fresh argument and by the end of 1969 only fifteen states had signed. Actually the clause was meant to help Udall save face among the conservationists. He had riddled it with loopholes. In one part, for instance, the clause says existing standards cannot be lowered unless "such change is justifiable as a result of necessary economic and social development," a statement which could mean anything to anybody. Still, the states are refusing to sign the non-degradation clause, and it has become the major obstacle to putting water-quality standards into operation.

Under Udall the water-pollution program drifted aimlessly. A great deal of the time was spent in helping President Johnson persuade the citizenry of the bold steps being taken to preserve the environment, when he was doing nothing. In fact, there was no money for pollution programs because the money was going into the Vietnam war.

Thus, after legislating reforms meant to strengthen the water-pollution program, the crusaders, Muskie and Blatnik, agreed to sinking the program into the Interior department, an act which very nearly ended this crusade.

4 | Fruits of Modern Technology

FINANCING SEWERS

As a practical matter the methods of treating sewage have not changed much since the turn of the century. The plants have changed shape a bit, and the processes speeded up, but the basic concepts remain as they were developed in England and at the Lawrence Experiment Station.

According to the Federal Water Quality Administration (FWQA), formerly the Federal Water Pollution Control Administration, the government agency which directs the national pollution program, more than 90 percent of all communities in the US are sewered and provided with some form of primary or secondary treatment. Primary treatment usually involves a sewage system which carries wastes from the home through a network of pipes to big tanks. There the sewage is allowed to settle for several hours. The liquid is poured off the top into a stream or lake. The solid matter, or sludge, on the bottom is carted away. Sometimes, depending on the place and the time of year, the liquid effluent is chlorinated as a precaution against disease, especially in instances where it flows out toward bathing beaches.

Primary treatment slows down pollution. It reduces the Biological Oxygen Demand (BOD) by one third. BOD is a measure of the strength of sewage in terms of the amount of oxygen required to sustain decomposition of the waste by bacteria. It is a common measurement used in gauging the amount of pollution from organic sources.

Bacteria in the water help cleanse streams and lakes by consuming pollutants. Along with fish and other living organisms, the bacteria need air, which they get from the atmosphere or from the life processes of aquatic plants. But the natural balance is thrown out of kilter when a large amount of sewage, crammed full of bacteria, enters a stream and increases the demand for dissolved oxygen to a point where the system cannot supply it. At that point, the system begins to break down.

In the past fifty years many larger cities, located along polluted waterways, have taken the additional step of

providing a secondary form of biological treatment. Now
the government wants all communities to adopt some
means of secondary treatment. This usually consists of
some sort of further biological treatment beyond the
settling stage. The most thorough means of secondary
treatment has been the trickling filter, as developed by
the Lawrence Experiment Station. Sewage is poured
over several layers of rocks or rock-like forms. The
bacteria adhere to the rocks, and feed on the sewage as
it comes through. But because trickling filters take up
much land, many cities adopt a form of activated-sludge
treatment instead. In an activated-sludge system, the
fluid goes from a settling tank into other tanks where it is
mixed with air and exposed to sludge alive with bacteria.
The secondary treatment breaks down the organic pollution load, reducing BOD by up to 90 percent.

After this process, the effluent is poured into a waterway. In some cases where communities border the ocean,
it is piped out to sea. Los Angeles runs an outflow pipe
seven miles out into the Pacific. In Miami, Florida,
where there is barely any sewage treatment, the sewage is
piped 7,000 feet into the ocean. It floats to the top of the
ocean and comes ashore five miles south of where it was
discharged. Other Florida communities follow this practice. The idea is to send the sewage into the Gulf Stream,
which can carry it far away from shore. Sewage, treated
or untreated, is pumped out to sea in many localities up
and down the Atlantic Coast while sanitary engineers
argue about whether or not it comes ashore.

The sludge left over in the bottom of the treatment

tanks is sometimes dumped at sea, where it causes blight on the ocean bottom. Over the years it has built up around pierheads of the Eastern port cities and is a serious problem in polluting the rivers and harbors. Chicago is experimenting with sending the sludge by pipeline for use as land fill in strip mine areas fifty miles away from the city. Some cities burn sludge which can cause air pollution. Milwaukee and Washington, D.C., sell sludge as a fertilizer base. Around the nation's capital, sludge is used to fertilize edges of parkways. But it is low-grade material and is considered hardly worth the effort.

Biological treatment has little effect on many pollutants. Inorganic pollutants such as metals or phosphorous pass unaffected through primary and secondary processes. Nitrates change their form but still emerge as nitrates. Bacteria do not affect pesticides, acids or salts; instead these wastes poison the bacteria and break down the whole treatment system. This is a common situation because so many municipal sewage plants also process industrial wastes. Of 280,000 manufacturing businesses in the US, all but 25,000 discharge into municipal sewers. The bacteria, which the sanitary engineers jocularly call "the bugs," sicken on occasion and the system slows down. Sometimes large amounts of industrial waste kill the bugs outright. New batches of bacteria must be grown and that can take from a week to a month. In Amarillo, Texas, a company which makes engravings for the local paper dumped zinc into one of the city's sewers, putting it out of commission for one month. During that time the sewage was bypassed raw into a creek bed. At another Amarillo plant, a bakery discharged 1,800 pounds of flour into the sewer killing the bugs and destroying the sewage plant. In Burbank, California, large amounts of chrome wastes from plating companies slow the Los Angeles sewage treatment system. In Phoenix, a General Electric computer plant has discharged chrome into the city sewers. In Seattle, Boeing has knocked out some sewers by chrome dumping. The Samsonite Corporation, in Denver, discharged chrome and poisoned that city's sewers. The company recently built a holding tank to reduce the chrome flow. In New York City, where officials have been struggling to make a big new sewage treatment plant at Newtown Creek work properly, they have been plagued by discharges of heavy grease from different industries. The grease coats the bugs and cuts down their efficiency. Rather than order the industries to quit discharging grease, the New York officials felt they had to cope with the waste, and devised a way to screen out the grease.

Many cities in the United States have combined sewers; that is, the sanitary pipes leading from dwellings feed into a sewer which also collects water and debris running off the street. This sewage is mixed together in the pipes and feeds into the central-treatment plant, which is only designed to handle a specified flow of sanitary sewage. Many sewage treatment facilities, built years ago, are already operating beyond their capacity.

In a heavy rain, the whole system can go out. The pipes fill and back up. Automatic valves open and bypass sewage directly into the nearest waterway. In New York City, for example, sewage-treatment facilities bypass once a week all year long. At the big plant on Ward's Island, when there is a heavy rain storm, the system backs up through interceptor pipes into the Bronx and upper East Side of Manhattan. Bypass levers are triggered and the sewage pours into the East River. In New York this process is not as simple as it sounds, for some sewer pipes are connected to the rivers at low-tide levels. Thus, at high tides the pipes are flooded with both river water and backed-up sewage. In Boston the sewage system bypasses once every five days. Combined sewers present an enormous problem. But to separate them would cost $15 billion. New York is experimenting with a lagoon where storm waters can be held for a while, then expelled into a bay after chlorination. Chicago wants to build deep caverns beneath the city to hold the water until it can be treated.

Because of the overloaded systems, sanitary engineers change methods of operation. What they often do is to shorten the time sewage stays in the treatment plant. That in turn reduces the amount of time bacteria has to work on raw sewage.

The job of running a sewage-treatment plant is tricky and hazardous. The plants are expensive, often intricate. The sanitary engineers, who man the controls, must forever be on the lookout for bad weather which can foul the works from overflow. It is in their interest to keep in close touch with industries on the sewer line. Friendly companies call up after someone has dropped toxic materials into the pipes, and the men at the sewage works can get the plant ready for the poison. But in Phoenix, two men were killed while working in a sewer when a company dumped in a load of cyanide.

Sewage plant operators sometimes work only half a day and hold other jobs. They often received little or no training. While the Federal government nominally requires that localities applying for sewage-works grants take care in the maintenance and operation of the plants, the rule is not strictly enforced. One study by the FWQA in 1962 showed that only one third of 970 plants inspected across the country maintained operating and laboratory records. About 40 percent of the plants were bypassing sewage. A more recent survey by the General Accounting Office showed that minimum criteria published by the states and the FWQA were not met in fifty-nine of the sixty-nine plants surveyed. In plants which the GAO inspectors actually visited, there were maintenance failures in eleven of twelve cases. In these instances, the GAO reported serious bypassing and breakdowns because of industrial poisons.

There is considerable research aimed at replacing the bacteria with a more reliable chemical means of sewage treatment. But so far the possibilities are too costly for any widespread use. The government policy meanwhile is to encourage communities and industries within a river basin to join together in regional sewer systems, and work out a scheme for disposing of sewage based on the assimilative capacity of the waterway. The model for this sort of scheme is the Delaware River Basin Commission which oversees the Delaware River, running through Pennsylvania, New York, New Jersey and Delaware, from the Appalachians to the Atlantic.

In 1961 the Commission was established to guide regional planning, development and management of the valley's water resources. The members include governors of New York, New Jersey, Pennsylvania and Delaware as well as the Secretary of the Interior. The Commission functions in part as a conciliation bureau for disputes among the states over access to water supplies in the headlands of the Delaware River valley. It attempts to work out arrangements for water diversion so that the flow of the river is not seriously reduced. The situation is especially difficult around Philadelphia; as the river flow declines, the salt tides, which usually halt just below the city, begin to creep up river, approaching the point where Philadelphia's drinking-water intakes come out of the river. The Commission is also sent to organize flood-prevention schemes and combat water pollution.

Nominally, the Commission has broad authority. Any organization which plans to discharge waste into the river must have construction plans approved by the Commission. It maintains surveillance teams to gather data on water quality, and the Commission can take action in both interstate and intrastate pollution situations. However, its activities have not amounted to much so far. While the Commission may have authority to act against polluters, in practice matters are left to the individual states. Surveillance of water quality is done mainly by the states. While the Commission has the power to grant construction licenses, it sometimes does not exercise that power. For instance, it stayed clear of Public Service Electric Company's proposed nuclear power plant at Artificial Island on the southern reaches of the Delaware. While the plant constituted a real danger because of radioactive and thermal pollution, the Commission paid little heed. The Commission has come out against the government's non-degradation standard for water quality. The non-degradation standard is meant to assure that water quality does not go below existing levels. Officials at the Commission frankly concede that while they intend to order pollution abatement, they are powerless to do anything unless towns and cities vote bond issues to pay for the work. One of the municipalities, the city of Philadelphia, is refusing to put into effect the Commission's proposed water standards on the grounds it does not have the money, and that the sewage ought properly to go into the river.

The Commission is best known for its plan to abate pollution on the Delaware Estuary. The Estuary is a length of river which runs eighty-six miles from Trenton, New Jersey, south to Liston, Delaware, where the stream widens and turns into the Delaware Bay. This is a heavily populated, industrial area, and the pollution in the Estuary is among the worst in the country. It is made especially severe because tons of sludge which have been poured into the river over the years from sewers at Philadelphia and Trenton have caked into large deposits on the river bottom. The tides make the emptying of the Estuary into the Bay a slow process.

The usual procedure in reducing pollution is to require the polluter to stop putting the offensive material into the stream. Under the Delaware River Basin scheme, the approach is different. Each industry or municipality which discharged into the river came forward with its own data on how much and what sort of material it was putting into the river. The Commission then determined, depending on the set of standards adopted, how much pollution could be tolerated in the river. Once it reached that figure, it allotted each discharger a certain permissible load of pollution. This proved to be an embarrassing situation, for it developed that some of the polluters had presented the Commission with inaccurate data. In hopes of getting off lightly, they made it appear as if their pollution load was small. Thus, they received a small allocation of permissible pollution. The result is that industrialists who originally endorsed the Commission are now denouncing it.

A study of the Estuary was made by the Public Health Service taking into account the various types and loads of pollution, as well as the types of industry, the expected rate of growth from industry, population figures, and other factors affecting pollution. Out of this, the government experts built a model which set forth five different sets of possibilities. If so much money were spent, then one set of conditions could be achieved; if more money was spent, then another type of conditions could be achieved. The idea was to permit people to balance the costs against different levels of improvement of the waterway.

The kind of information put together in the Delaware study is useful, but it also raises serious questions. The decisions as to how much pollution exists or should exist become the province of experts. The arguments and decisions are made by technical planners, in this case differences are argued out between the cost-efficiency government economists and engineers who want to build sewage plants and develop waterways. Pollution control becomes a process through which professional groups vie for political power.

Ecologists differ fundamentally from these modern technologists in that they do not share the same perceptions of the political economy. For them, cost effectiveness, economic benefits or industrial growth are of secondary value. They believe the most important thing is to preserve the natural order. Man must accommodate himself to the planet, not, as E. B. White said, "beat it into submission." As ecologists begin to act, putting their ideas into practice, they emerge as radicals. They are anathema to the interests which now control the politics of pollution.

To understand a little of what ecologists propose in

the way of a sewer system, it is interesting to look at recent experiments at Pennsylvania State University. They were carried out under Louis T. Kardos, an environmental scientist at the Institute for Research on Land and Water Resources. Kardos' inclinations are reminiscent of Chadwick.

Pennsylvania State University and the nearby town of State College dominate the Nittany Valley of Pennsylvania. Sewage from both town and university is usually collected, given primary treatment to settle out the solids, and exposed to bacteria to treat what remained. The effluent, containing phosphates and nitrates, was poured into a small stream which flowed into a larger stream. Both those streams were choked with algae, and the trout population was replaced by suckers. As the water ran out of the area and down the streams the ground-water levels were depleted.

A team of scientists at the university went to work on these different problems in the early 1960s. Their aim was to return the minerals in the sewage to the land, restore the water to the ground-water reservoir, reduce the over-fertilization of the creeks and eliminate the need for adding chemical fertilizer to crop land. Their scheme seemed simple enough. They attached a pump to the university-town sewage plant, and piped sewage, which had already received secondary treatment and been chlorinated, back through the forest land. At intervals sprinklers spread the effluent over experimental plots in both agricultural and forested land. While the sewage water had been treated, it was not potable, and

the idea was to see if by letting it trickle through the
undergrowth and into the earth it would become
drinkable. The scientists also wanted to find out what
effect it would have on the growth of trees and on the
amount of corn, hay and oats produced in experimental
plots. The experiments were made with care and ran
from 1963 through 1968.

The effect of the waste water was to increase the crop.
Plots treated with fertilizer produced two and a quarter
tons of alfalfa hay per acre while plots sprinkled with
one inch of waste water each week yielded double that
amount. Increasing the amount of waste water sprinkled
on the crops further increased the crop yields.

A comparison was made between plots irrigated with
well water and fertilized, and plots irrigated with waste
water, but not fertilized. The waste-water irrigation resulted in higher yields—116 bushels of corn per acre
versus 107 bushels. The experiment suggested that
sewage effluent can take the place of both commercial
fertilizer and the usual irrigation water. In plots where
sewage water was used to irrigate forest lands, stands of
white spruce nearly doubled their length over five
years. The ground vegetation also was more abundant in
the plots irrigated with sewage water.

Samples of water were taken as it trickled down
through various levels of the soil. Phosphorous was reduced by 99 percent in the forest lands before it reached
the one-foot level during the first year. In succeeding
years the ability of the soil to remove phosphorous
declined slightly. Between 85 and 92 percent were re-

moved. In agricultural lands, the amount of phosphorous removed ranged from 22 percent in corn silage
irrigated with two inches a week to 63 percent in red
clover irrigated with an inch. Samples taken from deep
and shallow wells at the site of the experiment showed a
phosphorous level of 0.04 milligrams per liter of water.
This was about the same amount of phosphorous as
was found in the creek above the sewage outfall.

Both phosphorous and nitrogen are believed to be a
major cause of the overgrowth of algae. In describing
the Penn State experiment in the March 1970 issue of
Environment magazine, Kardos describes the experiment as it relates to nitrogen: "Since 1965, there has
been no further increase in the concentration of nitrates
in the percolating soil water in any of the forested plots
receiving one-inch- or two-inch-per-week applications.
On the control plots, the nitrate-nitrogen concentration
has remained less than 0.5 milligram per liter. Where
four inches of effluent have been applied weekly, nitrate-
nitrogen levels have approached ten milligrams per liter,
the USPHS limit for potable water.

"In the agricultural plots, crops frequently removed
a high percentage of the nitrogen added in the waste
water. In fact, the amount of nitrogen in the crops sometimes *exceeded* that added in the waste water, showing
that the plants could utilize more nitrogen than was
added and were drawing upon the nitrogen already
present in the soil from decaying organic matter and
previous treatments with chemical fertilizer."

Both nitrogen and phosphorous were removed from

the system in large amounts when crops like corn and hay were harvested. A 111-bushel crop of corn removed 112 pounds of nitrogen, 41 pounds of phosphorous, and 39 pounds of potassium. These amounts were equivalent to 105 percent of the nitrogen.

As the State College area gained population the ground-water table dropped. In part that was for natural reasons because of a deficit in rain. But it was also due to the fact that the university took 2 million gallons of water a day out of the creeks. And instead of returning the water to the valley, the waste water was discharged downstream, where it ran out of the area. Between May 1962, and May 1965, the ground-water table dropped sixty feet. The water table below the irrigated area also dropped, but not so much. It declined twenty feet during the same period, which suggested the value of returning water to the valley. In the 1968 *Yearbook of Agriculture,* Kardos and two associates—William E. Sopper and Earl A. Myers—write, "Another benefit from using the farm and forest land as a living filter occurs as the cleaned waste water trickles down to help fill the ground water reservoir. Ultimately, this water, which now meets the US Public Health Service standards for drinking water, works its way back to the streams or lakes or to the wells and springs of the region. Data at the Penn State project during a dry year indicated 80 to 90 percent of a weekly application of 2 inches of waste water was recharged to the ground water. During a normal or wet year, even larger amounts of water would be added to the ground water reservoir."

While prospects for more of this sort of treatment looked promising, the local sewermen worked for an opposite goal. The State College area is growing fast, changing from a rural to suburban area. A new sewage plant is being built several miles below the existing plant. It will handle the new suburban areas. The new plant will have both primary and secondary treatment, but that may not be sufficient. There are plans to make a man-made lake below both sewage plants, and thus the treatment facilities will pour their loads of nitrates and phosphates into this cul-de-sac, in all likelihood breaking down the natural system and causing the growth of algae. If that happens, the state agency proposes to move to some form of chemical method for removal of wastes. No one seriously considers the idea of using the land. The state health department is against the idea, for it fears nitrates sinking into the ground soil will cause a danger to drinking-water supplies, although the experiment suggested this would not be the case. Local and state officials are sticking to their policy of building more sewage treatment plants which will further deplete the resources of the valley by washing them down stream, where they will cause more pollution.

The irrigation level of two inches a week, which seemed promising in the Penn State project, showed that disposal of one million gallons of waste water per day would require only 129 acres of land. To dispose of similar waste waters from a city of 100,000 people would require 1,290 acres or about two square miles. The productivity of farm and forest land on these areas might will be improved. Kardos in the *Yearbook* article says, "Since any disposal system must operate throughout the year, in northern climates when the temperatures are low, the system must rely more on the adsorptive capacity of the soil and less on the microbes and roots. Forested areas fit into the system well by providing better winter infiltration conditions and larger phosphorous adsorptive capacities under the acid conditions associated with forest soils. Combinations of cropland and forested areas will provide the greatest flexibility in the living filter concept."

And Kardos concludes in *Environment*, "The living filter concept suggests that it might be well in many cases to re-evaluate the present trend to even larger and more centralized sewage-treatment facilities, built on the handiest waterway, and consider decentralized facilities, convenient to the nearest effective living filter."

BOONDOGGLES

Industrial sewage presents the most serious pollution problem, both because of its content, which bacteria cannot break down and because of the amount—for every one pound of BOD from a person there are four pounds from industry.

Although many industries discharge their wastes directly into municipal sewers, they infrequently pay much, if anything for this service. According to surveys of the Federal Water Quality Administration, about

85 percent of communities charge some sort of a fee for use of sewers. About half of those communities levy an additional charge on industry for handling its wastes. In the northeast, where manufacturing is concentrated and historically pollution has been the most severe, user charges to industry are rare. In Boston, for example, industry and residential users pay the same rate for sewerage, and that rate is based on the amount of water used. Los Angeles only recently added a surcharge for industrial users. There is no additional charge for industry in San Francisco. In New York City, charges to industry depend on the volume and strength of the sewage.

Thus, while the general citizenry pays for the sewerage system, industry takes advantage of it. The system answers the needs of other economic interests, as well.

To get a better idea of how a sewage system works, it is useful to look at one system in some detail. A 1965 study by Francis X. Tannian, then a student at the University of Virginia, makes it possible to learn a good deal about the way sewers are built and operated in the Washington, D.C., suburbs. Tannian examined in detail the workings of the Washington Suburban Sanitary Commission. The Commission oversees the water and sewerage systems for Montgomery and Prince Georges counties, the two mushrooming suburban counties adjacent to the capital city in Maryland. The members of the Commission are appointed by the governor of Maryland, and they are in charge of three treatment plants and many miles of sewer pipes. It is an important political job because sewers are requisite to real-estate development, a booming business in the Washington suburbs.

Over 80 percent of the Commission's assets go for equipment which transports the sewage. During four years ending in 1963, $25 million was invested in the sewage system, and of that total, $17.5 million went for pipes to cover the great distances involved. Both counties grew rapidly out from the city, and real-estate developers leap-frogged across old farms. The sanitation commission followed to provide service.

Almost all the users of the system (98 percent) pay for sewerage on the basis of amounts of water they use. In 1963, metered water customers paid 25 cents per 1,000 gallons for sewage collection and treatment. This is important to understand because the costs of pumping, treating and piping sewage are not remotely related to the amount of water used.

One key cost is the length of pipe. The flat-user charge, which most people pay, does not account for all those who really benefit from the sewer. For instance, a development far distant from the treatment plant does not pay more for treatment than does a house or development a couple of miles away, even though the cost of running the pipe to that development is enormous. As it now operates, the sewer system contains a hidden subsidy, an incentive for land speculation. If, alternatively, user costs were increased the farther away one went from a treatment plant, such rates would weaken speculative land practices and the construction business.

Other costs to the Commission also depend on the length of lines. Maintenance costs depend on the amount of area covered. The more joints there are, the more likely it is that the system will break down.

The Commission supports the construction business in another way: stoppages occur at the upper ends of sewer systems where the rate of flow is low because of the few dwellings; then, in excavating, construction companies pour mud into new sewers, clogging them.

The system is inequitable in another respect. The Sanitary Commission charges nothing extra to users of domestic garbage grinders, which place extra cost on the plant, nor does it ask business which put foam or other detergents into the system to pay any additional amounts.

Residents who use a lot of sprinkling water, whether or not the water ever goes into the sewer main, pay more for the sewage because the water rate is tied to the sewage rate. On the other hand, while the Commission does not advertise the policy, large-water consumers, such as bottling companies, can apply for a separate metered sewage connection and they need pay only on the basis of the flow into the sewers.

The Washington Suburban Sanitary Commission is able to boost revenues for its sewage system, by advertising low water rates. They hope that by selling more water, income for sewers will rise accordingly. Over a long term, this sort of policy is disastrous. Increasing the use of water in a borderline drought area reduces the ground-water supplies and cuts down the flow of the Potomac River from which much of the drinking water comes. Where the drinking water is withdrawn, sewage is pumped back into the river, changing the ecology.

Most owners of property who apply for the first time for sewage must pay a front-use fee, which supposedly covers the cost of attaching the sewer. But public properties, charitable institutions and agricultural properties are exempted from the front-use charge. Land speculators frequently call their holdings "farms," in order to persuade the Commission to view their land as small acreage, and thereby avoid paying the front-use fee.

The Commission has covered the cost of new sewer lines through a policy of accepting "contributions." In the past this has had a curious effect. In 1962, the International Business Machines Corporation announced it would build a new plant in Montgomery County and purchased land zoned for low-density housing seven miles beyond the limits of the sanitation Commission's territory. IBM had no difficulty in having the land rezoned from low-priced, low-density land to the highest-priced industrial category. In violation of its own master plan the sanitation Commission agreed to build the seven additional miles of sewer pipe to accommodate the company's new plant. IBM paid $350,000 as a "contribution" toward the cost of laying the lines. The sanitation Commission customers bore the remaining $950,000 in costs. Shortly after IBM announced its move, and the land was re-zoned, plans were announced for a new shopping center and industrial park. The Commission agreed to service these developments. In this case, the industrial park paid $150,000 of the $1 million cost. All in all the Commission laid twelve miles of sewer line into the countryside across vacant lands to connect the system to IBM and a new shopping and industrial center. The two big users paid $500,000 to obtain a sewage system which cost $2.4 million.

In May 1961, the chief engineers of the Commission rejected sewer lines for development west of the town of Olney in Montgomery County. The Commission claimed development of the Olney region might result in silt deposits in the Patuxent River, a source of the county's water supply, and hence lead to increased cost in bringing in water. Two years later that decision was reversed. The applicant for the new sewer was A. W. Turner. His chief counsel was M. B. Wheeler, a sanitary commissioner. Turner paid no contributions. Wheeler argued before the county re-zoning board in behalf of Turner. It then developed that Wheeler owned land in the path of the sewer. A year later the zoning commission gave up the fight, and Turner won his right to re-zone the land.

In recent years the Washington suburbs have sprawled down both sides of the Potomac River. This is lush territory for the realty developers and they have been busy ripping up the place, building suburban housing all over. In Prince Georges County on the Maryland side of the river, developers were permitted to build septic tanks. The Commission in several cases did not check septic-tank drains carefully; citizens discovered they were hooked into storm gutters, and did not lead to proper drains in the ground. The raw sewage was turning up in public places. A grand jury eventually indicted several public-health officials for failing to make proper connections. Those indictments were subsequently dropped. The growing population resulted in a new sewage plant being built at Piscataway Bay, off the Potomac. The Commission insisted the plant could handle all the sewage in the area, but in fact it was designed for a relatively small capacity, far below the amount required. In case of overflow, the Commission arranged with the city of Washington to bypass the plant, and run the sewage into the Washington system. Washington sewage is treated at a large works at Blue Plains, also on the Potomac. What happened was that the Piscataway plant could not handle the load and the sewage was shunted to Washington. The Washington system is already clogged and receiving insufficient treatment. In this way the Suburban Sanitary Commission piggy-backed its sewage onto the residents of the capital.

While the precise situation may differ from place to place, charges for sewage usually follow the above pattern. Not only are the pricing policies inequitable, but by subsidizing real-estate developers, they are rigged to create more pollution.

Much of the government's anti-pollution legislation is written to benefit Wall Street, which hopes for an increasing market in underwriting sewer bonds. Although it would make most sense for the government to finance new sewers itself, either through direct appropriations or by issuing bonds itself, thus far money is shunted to communities in order to support the burgeoning tax-exempt municipal-bond market, where sewer issues are sold.

Since 1956, the government has made available limited amounts of money in grants to local governments to help them pay for the construction of sewage-treatment plants. The total spent never was higher than $200 million a year until under Congressional pressure, Nixon agreed to spend the entire $800 million appropriated in 1969. The money is spread around the nation according to a complicated formula which takes into account geography and population. In practice, the formula has worked against big cities where sewage-disposal problems are most acute. Since the projected costs for pollution abatement in New York state alone are over $1 billion, the program does not appear substantial.

For some years, in both the Johnson and Nixon Administrations, bureaucrats haggled among themselves to change the grants program, so that as the public clamor over pollution rose, there was at least a semblance of a sewer-building program. In the Johnson Administration, teams of officials from the Treasury, Budget Bureau, FWQA, White House and Congress met to work out some new system of financing.

Essentially the problem was this: the grants program ties the government to a direct annual-expenditure program, which is likely to get much bigger. The system

provides constant annoyance to states since it has the
Federal government intrude directly into the affairs of
municipalities.

The idea was to create a financing scheme where the
government could either underwrite or guarantee sewer
bonds. The guaranteed bonds could be lucratively sold
in the public markets. Another plan was to create an
environmental authority; it would buy sewer bonds
from communities at below-market interest rates, then
sell them over the counter to the general public. (This
is imitative of the Federal National Mortgage Association which sells housing bonds.)

The basic problem comes down to the investment-
banking community. In recent years the private investment banks have become very much interested in assist-
ing communities in financing sewers. In general, the
investment bankers have argued strongly within the
government for more Federal financial support to communities to build sewers. Sewer bonds are generally
sold in the tax-exempt municipal bond markets to
wealthy individuals.

In 1967, the Federal Water Quality Administration
contracted with Eastman Dillon, Union Securities & Co.
to make a lengthy study of sewer financing; it was conducted by James Lopp II, a partner in the firm. For assistance, Lopp brought in John Mitchell, of Nixon,
Mudge, now Nixon's Attorney General, then a well-
known bond lawyer. Eastman Dillon, as it turned out,
became the financial consultant to the Interior Department, and President Johnson named Lopp as a member

of the Federal Water Quality Administration. Finally, Joe G. Moore, the former Texas water commissioner who had served under LBJ as commissioner for water pollution, quit his job at Interior in 1967 and became a partner at Eastman Dillon.

Eastman Dillon saw two possibilities in the burgeoning sewer business: for one thing the firm could become a financial advisor to communities with pollution problems, and draw up their financing plans. Since those plans often depended on re-working political configurations, Eastman Dillon advocated regional sewers, lumping communities together in a common sewer system. Industries would participate in the regional sewers. Eastman Dillon is a major underwriter of corporate bonds, and its corporate clients are concerned about financing pollution-abatement schemes. If it could take a hand in organizing regional sewer systems, Eastman Dillon would have something tangible to offer both corporate and municipal clients. In effect, the company would brokerage their interests.

For the investment banking concerns it is imperative for the municipal bond market to expand in an orderly fashion, and for that to happen, the government would have to keep on spending more money to back sewer bonds. Not surprisingly, the thrust of the Eastman Dillon report was to increase Federal spending in such a way as to boost the tax-exempt bond market.

This puts the private investment banks at odds with the Treasury. The Treasury historically takes a dim view of the tax-exempt bond markets on the grounds that they offer wealthy, institutional investors a tax shelter. During the Johnson Administration, officials within the government wrangled over alternative financing schemes but could never agree among themselves. So the policy stayed unchanged.

Shortly before Christmas 1969, the Secretary of Interior, Walter Hickel, suggested a new idea. Under his plan the government would abandon the grants program, and instead local communities would be encouraged to sell bonds on the open market to pay for sewage-treatment facilities. These bonds would be popular because the government would pay the principle, leaving the communities to pay the interest. In effect, the communities would pay half and the goverment half. That is pretty much how the grants program worked, but supposedly the Hickel plan would put the government's leverage in the money markets to more innovative use, and the number of communities building sewers would increase. If this plan worked, communities would flood the bond markets, driving up interest rates. Needless to say, there was considerable enthusiasm for the plan among investment bankers.

To them it must have seemed a sign of ultimate victory. For during the tax-reform debates of late 1969, the Treasury and the Congressional leadership fought to do away with tax-exempt municipal bonds altogether, thereby shrinking or eliminating that market.

John Mitchell, the former municipal bond lawyer and Attorney General, was generally credited with helping persuade Nixon to save tax-exempt bonds. It seemed as if the market for tax-exempts would increase, not diminish.

The details of the Hickel plan were never officially announced, but, as they were sketched out behind the scenes, it became clear that for several reasons the plan would actually diminish the number of sewers built: in a period of high interest rates, it would discourage many localities, already heavily in debt for schools, municipal services, police, etc., from going to the bond markets, and consequently it would reduce the demand on the government for new sewers. That, in fact, may have been the idea.

The Hickel plan never went anywhere. In his message on the environment in 1970, Nixon proposed another method of financing. It was endorsed by both Democrats and Republicans. Under the proposed setup, the government would continue to provide grants-in-aid to communities up to 55 percent of the cost of the major components of the sewage system (that is, waste treatment plant and interceptor pipes). If communities still could not sell their bonds in the public markets, a new governmental agency called the Environmental Financing Authority could step in and buy the community bonds at interest rates equivalent to those on the tax-exempt markets. The Environmental Financing Authority, in turn, would sell its own bonds publicly, and they would be taxable.

This aroused the enmity of the investment bankers who saw it as part of the Treasury's continuing effort to ruin their business in tax-exempt municipal bonds. In addition, the National League of Cities, which reflects the interests of many large cities, was sceptical of the idea. It pointed out that communities had limits of indebtedness beyond which they could not go, and that interest rates were at all-time highs. It was difficult to see how the government proposal would cut around either of these obstacles. Also, as the proposal was written, the water-pollution agency would gain substantial political power, for the agency would certify whether or not the community would be eligible for Federal funding; the Secretary of the Treasury would determine the interest rate paid by the community.

Not only are there indications that the economics of the sewer business are distorted for profit and boondoggle, but there is strong evidence to show that the government's expenditures for sewage-treatment facilities were wasted.

Between 1957 and 1969 the FWQA awarded grants of about $1.2 billion for construction of more than 9,400 projects which cost $5.4 billion. While the new sewers may have helped slow the spread of pollution in a general sense, over the short run, the government's sewer grant program created chaos and helped make the fight against pollution more difficult than ever. A study made for Congress by the General Accounting Office describes case by case how the government money was used. A few of the more celebrated examples follow:

The Nashua River, which flows through fifty miles

of Massachusetts and then enters New Hampshire, has been formally polluted since 1936, and in 1967 the FWQA called the stream "one of the most disgusting rivers in the country." Between 1961 and 1965, the towns of Leominster and Ayer, Massachusetts, both of which discharge waste into the Nashua, were awarded $700,000 in Federal grants for construction of sewage projects. The resulting improvements were insignificant because, as the government well knew, 80 percent of the BOD discharged into the river comes from industrial sources which were clustered *upstream* from Ayer and Leominster at Fitchburg. The two biggest polluters at Fitchburg are Weyerhaeuser Paper Company and Fitchburg Paper. During the summer, the Nashua has a reduced flow, totaling about 8 million gallons a day at Fitchburg. During that period, the entire river flow goes through both the Weyerhaeuser Paper Company and Fitchburg Paper and then is expelled back into the river. At that point, the Nashua, practically an open sewer, flows on downstream, past Ayer and Leominster.

Fitchburg's own sewage system is outmoded, and, according to a state plan, it is meant to construct a facility costing $2 million. Under that plan the industries were expected to make their own improvements. However, on looking into this business, the GAO auditors discovered Fitchburg had changed its plans around, and proposed to build two separate waste-treatment plants for a total cost of $17 million. The planned projects would treat both industrial and municipal wastes. Both plants would be eligible for 50 percent grants from the FWQA. In this way, of course, the expense of cleaning up the river was shifted to the local citizenry who will bear the burden of paying off the sewer bonds ($15 million in excess of what was originally planned), and to the general tax-payers, who will pay for the program through the Federal grants program.

The Tualatin River in Oregon is a sluggish stream which naturally has a very low flow in the summer. The flow is further depleted by diversion of the water for irrigation in neighboring fields. The river is in bad shape because effluent from sewage plants depletes the dissolved oxygen and adds nutrients. They result in the growth of slime and algae. In the twelve years between 1957 and 1969, twenty-one Federal grants totaling $2.6 million were made to municipalities on the Tualatin to help them in building or enlarging thirteen treatment plants, interceptor sewers and outfalls.

In 1956, local officials were concerned by stunted development along the river because of inadequate sewage facilities. They hired an engineering firm to prepare a master plan for sewage treatment. The plan, completed a year later, said that the creeks which form the Tualatin's tributaries should not be used except as temporary channels for sewage effluent because most of them were dry in the summer. The plan suggested that instead of small, local treatment plants, communities should band together and build interceptors to carry the sewage out of the area to a large new plant on the

Willamette River. The total cost for the interceptors and the plant would be $20 million. Three counties took up the plan, and it was adopted by the state.

However, the FWQA never recognized the plan. By 1962 nothing had happened, and the municipalities along the Tualatin began to doubt anything would ever happen. They turned to the FWQA, which provided them with funds to enlarge the sewage plants along the dry tributaries, in effect creating a lovely system of open sewers all year round. The situation worsened and the state put a stop to building new plants unless they could avoid discharging sewage into the river at periods of low flow. In 1969, another master plan was prepared, and like the original master it too called for a regional system. So, with the exception of its open-sewer system, the Tualatin counties were right back where they started in 1956.

The GAO auditors tried to find out why the FWQA gave out grants to small treatment facilities which drained into the dry creek beds. (Ten of twenty-one grants made in the river basin were for projects which discharged into the dry-river tributaries.) The GAO looked into three grants with special care. They cost the government $451,000. In order to qualify for a grant, a community must demonstrate it is part of a comprehensive program. According to the FWQA, the three plants in question qualified because they appeared in a list published in the Federal Register. It was a list of municipalities, arranged in alphabetical order, which identifies places in need of sewage treatment. In the grants application, the three projects were certified by the state agency as being in "conformity with the state water pollution control plan submitted pursuant to section 5 of the Federal Water Pollution Control Act." The three plants discharged into the tributary even though the master plan adopted by the state said the tributary could not assimilate the treated waste water.

The FWQA insisted its financing methods had changed since the Tualatin fiasco, and that now approval for sewage plants is based on the projects' compliance with a comprehensive report dated January 1967. In 1969, however, the FWQA awarded a grant to build an interceptor which served a plant discharging into one of the Tualatin's dry tributaries.

The Pearl River, which flows on the boundary line of the Mississippi and Louisiana, is polluted, and two enforcement conferences considered the situation in 1963 and again in 1968. A 1962 survey, prepared for the first conference, said that two firms—Crown Zellerbach and Crosby Chemical Company—accounted for the major sources of BOD in the southern reaches of the Pearl. The survey also said the two companies along with two municipalities, Bogalusa, Louisiana, and Picayune, Mississippi, were the major sources of coliform bacteria. The government subsequently gave grants of $640,000 for sewage-treatment facilities to the two towns. The plant at Bogalusa removes about 4,500 PE (population equivalent) of BOD. But Crown Zellerbach, which is located in the area, discharges 400,000 PE of BOD. The plant at Picayune removes about 8,000 PE of BOD. But Crosby Chemical discharges 125,000 PE of BOD. As the towns struggle to make a clean-up, the industries put more and more filth into the water. Their discharge is of such a magnitude that it wipes out whatever progress is made by the towns.

The Willamette River serves two thirds of Oregon's population. Since 1965, more than 75 percent of construction grants provided by the FWQA to the state, or $10 million, have been used for construction in the Willamette basin. According to a 1967 FWQA report, 75 percent of wastes generated in the basin were from the pulp and paper industry, 14 percent from the food-processing industry and 7 percent from municipal sources. The water-pollution agency said that one of the most serious pollution problems in the country occurred on the lower stretches of the Willamette. According to the report, two paper companies—Crown Zellerbach and Publishers' Paper—both located on the lower stretches of the river, generated about one third of the wastes of the entire river basin. Most of the time, the two mills accounted for 99 percent of the waste load on the lower reaches of the Willamette. In 1960, the state cited the lower reaches of the river as unfit for various uses because of the low-dissolved oxygen and high-coliform bacteria counts. Four cities were directed to build sewage-treatment plants. That resulted in a reduction of 20,000 PE of BOD, an insignificant improvement when compared to the two paper mills, which continued to pour out their wastes totaling 500,000 PE of BOD.

On a 170-mile strip of the Mississippi River in Louisiana, from St. Francisville to Chalmette, the government spent $7.7 million between 1957 and 1969 to build treatment plants for six municipalities. As a result, pollution was reduced from 1,032,000 PE of BOD to 885,000 PE.

Although the river is the source of water supply for various communities, state officials told the GAO they were not concerned about formal BOD counts, claiming that the assimilative capacity of the river was sufficient to receive the waste discharges without adversely affecting the existing water uses. Eighty industrial plants discharge into the river. Included among them are forty-nine oil refineries and chemical plants and nine sugar refineries. The state operates a permit system for industrial-waste discharges, and it conducts a sampling program to test for discharge of toxic and other wastes that might interfere with water use. On looking into the state files, the GAO found the information on industrial discharges was scant. Information on BOD was contained on thirty companies, and information on volume of waste was available for fifty-two plants. But in the case of twenty-two of the fifty-two plants, there was not sufficient information to ascertain the nature or the volume of the wastes discharged. The GAO auditors say,

"We were advised by a state official that many of the permits did not contain information on BOD because either the types of wastes involved did not contain BOD or BOD effects of the wastes were not known. However, the data on BOD that was available showed that the BOD contained in the wastes, which were permitted to be discharged into the river, increased from about 970,-000 PE in 1957 to about 3,383,000 in 1967."

Thus the government spent $7.7 million to achieve a reduction of 147,000 PE in municipal sewage, while industrial pollution increased 2.4 million PE.

Since 1961, the FWQA has been attempting to help the towns of Saco and Biddeford, Maine, clean up the lower stretches of the Saco River by giving them both grants for municipal-sewage facilities. However, since most of the BOD comes from two industries—Saco Tannery and the Pepperell Manufacturing Company, these measures are inconsequential. Recently Saco announced plans to treat the tannery's wastes and put in to the Federal government for the money to clean up industrial sewage.

On the Ten Mile River, which runs between Massachusetts and Rhode Island, municipalities have built sewage-treatment plants which reduce domestic sewage. But these improvements are more than offset by industrial sewage flows which continue unabated. Industry accounts for two thirds of the 30,000 PE of BOD. Enforcement conferences were held in January 1965, and again in May 1968. The 1968 conference showed that

about half the industries were behind schedule in providing waste clean-up. Originally, industry used municipal sewers; the municipalities put a stop to this because
the industrial wastes killed the bugs. Since the companies could not shovel their wastes into the municipal
system, they did nothing at all. Finally, on the Presque
Isle Stream, which flows from Maine into Canada, the
state of Maine declared the water to be suitable for
bathing, recreation, fish habitat and water supply. In
1960 the state required the town of Mars Hill to build a
primary-treatment facility, which was partially financed
through the FWQA and accelerated public-works program. There was a reduction of BOD as a result of the
plant. Then, upstream from the Mars Hill, at Easton,
pollution became a problem. Fred Vahlsing built a
Vahlsing, Inc., plant at Easton. That plant makes frozen
french-fried potatoes. He also constructed a sugar-beet
refinery called Maine Sugar Industries. In that endeavor,
he received help from Senator Muskie. Two government agencies, the Area Redevelopment Administration and the Economic Development Administration
made loans to help expand both plants. In 1967 the
FWQA issued a report which showed the PE of BOD
from the food-processing plants amounted to 400,000,
while the total from Mars Hill was about 900. The
Vahlsing plants tried getting rid of their wastes in different ways. They sprayed it on fields, but in the winter it
froze. They put in it a dike, but the dike burst and
flooded the stream. Finally, they sprayed it in a swamp,
but the swamp overflowed into the stream. In 1969, the

GAO auditors were informed by state officials that the food-processing plants no longer planned to build treatment facilities. He said the industrial waste loads had increased far beyond the permit loads, to the extent that 90 percent removal of BOD would not permit the stream to assimilate the effluent.

The recitation of instances from the GAO study is not necessarily meant to demonstrate the severity of water pollution, although it certainly makes that point as well. But the GAO study demonstrates precisely how the Federal program functions to cover up the severity of the pollution situation. Again and again the concentration is on improving municipal sewer systems, which account for relatively small amounts of pollution, and allows the industries to continue on dumping in their filth. Or, in the more progressive schemes, industry persuades the municipality to take on its sewage on industry's terms. I have already indicated how difficult it is for sewage plants to cope with industrial sewage. Introducing certain kinds of industrial sewage into a municipal waste-treatment plant threatens to break down the whole system. In addition, the imposition of the cost of cleaning up industrial waste, laid on top of an already rigged cost system, charges the local citizenry with higher and higher rates for handling sewage for which they are not really responsible.

5 | The New Fuels Trusts

CORNERING THE ENERGY MARKETS

The world oil markets are in the midst of a terrific boom. Production and consumption of oil has sharply increased since the Second World War, and now amounts to 38 million barrels per day; it is expected to double by 1980. While there will continue to be a steady growth in Europe, the biggest jump will be Japan, where demand is expected to increase from 3.2 million barrels a day in 1969 to 6.1 million in 1975. The oil business has grown faster than the industry itself expected, in part because world production of coal has fallen off and because nuclear power for use in industry has run into high costs and delays.

Seven major oil companies dominate the oil business. They are Standard Oil of New Jersey, still the mother of trusts; Shell, Gulf, Texaco, Standard Oil of California, Socony Mobil and British Petroleum. Their percentage of the business has dropped from 90 percent in 1955 to 70 percent in 1969. However, it is not expected to go any lower. Privately owned independent companies have only 18 percent of the world market, and government-owned enterprises have about 12 percent.

Over half the known oil reserves are in the Middle East. Another 10 percent are in North and West Africa. Discoveries in this area have outpaced those in the Western hemisphere and in East Africa because costs in the Middle East and Africa are well below those in South America or Indonesia. With the construction of ocean-going super tankers, the cost of transporting oil to the new markets in Japan has declined.

Increasingly, oil companies think of themselves as "energy" companies. They see the future in terms of playing one energy source against another for maximum profit. They are also actively involved in the chemicals, fertilizer and plastics markets. So the corporate future of an oil company is very often directly tied to both expansion of the energy markets and an expanding business in consumer goods.

While the petroleum companies increasingly have come to control other forms of energy production, the money is in oil. So far the big oil companies have been able to control the industry from well-head to consumer. The Connally "hot-oil" act passed in the 1930s by Congress gives the states power to set levels for oil production. The authority is vested in state commissions dominated by large companies. The industry pretends the producing end of the oil business is competitive because there are thousands of small operators. In fact, two thirds of the crude comes from the large companies.

To protect the US oil market from a deluge of cheap, foreign oil, the oil men persuaded President Eisenhower to establish a system of import quotas, allegedly in the interests of encouraging oil exploration at home. Under the program, US refineries get "tickets" allowing for the importation of a certain amount of oil each year. The quota is based on their production, which means that the subsidy (foreign oil is $1.25 a barrel cheaper than domestic crude) goes to the large companies, not the small ones. Since the quota is given to refineries, the small producer, whom the system is meant to encourage, gets little help.

But the crux of the oil game is the tax system. The breaks are all loaded in favor of the producer. Until 1969, the producer could claim up to 27.5 percent depletion allowance on up to 50 percent of his income, which meant he could write off the cost of a well long after it was paid. (The tax-reform bill of 1969 pared the rate of depletion allowance to 23 percent.)

In addition, the producer can write off in one year certain "intangible" costs, including labor, materials, fuel and equipment repairs; practically all his major business expenses. Certain expenses incurred in foreign operations can also be written off. The results of the tax system are striking: Standard Oil of New Jersey showed a net income of $2 billion in 1967 but paid taxes of only 7.9 percent, or $166 million. Texaco's tax rate was 1.9 percent; Mobil's, 4.5 percent; Gulf's, 7.8 percent; Standard of California, 1.2 percent. Atlantic Richfield, one of the two companies which made the recent Alaska find, reported a net income of $145.2 million in 1967. The company paid no tax at all.

The tax system helps to explain a good deal about the structure of the business and the frenzy of its politics. Because the special treatment occurs at the producing stage, the big oil companies want to keep the price of crude oil high so as to facilitate a big write-off. The tax system, which at least nominally was meant to encourage competition, and hence in anti-trust mythology, to lower prices, in this instance works in reverse, and appears to create higher prices.

The over-all effect of the tax setup is to bring on a great glut of oil, and with it an enormous pollution problem.

The oil boom has led in turn to a boom in the tanker business. Half of all cargo shipped at sea goes by tanker, and with the oil men expanding their operations throughout the world, there is an increased demand for ships to shuttle the supplies from oil fields to new markets. The already congested shipping channels are becoming seriously crowded and accident rates at sea are on the rise. (Some 200,000 tons of cargo were lost at sea during 1948, compared with 500,000 tons in 1963.) When one of the new enormous tankers with 200,000 tons of oil aboard goes aground or rams into another ship, it is no ordinary accident. Tons of oil spilling from one of these gashed hulls can and have played havoc with the environment.

The world tanker fleet now numbers about 4,000 ships, and it is fast increasing in both numbers and size. The *Torrey Canyon,* the ship which broke in half off Land's End, England, in 1967, carried 119,000 tons of crude oil. Now it is common for super tankers to carry 200,000 tons. Tankers of 500,000 tons are being designed, and a one-million-ton tanker is under consideration. A survey by Davies & Newman, London tanker brokers, shows the world tanker fleet will increase in tonnage by 50 percent between 1970 and 1974. At the beginning of 1970, 491 tankers totaling 70.9 million deadweight tons were on shipyard order books. Of the new tankers ordered, 205 are of the 200,000-ton capacity.

About one third of the fleet is owned outright by the oil companies themselves, but through various devices they actually control many more ships. Big independent operators include Daniel Ludwig, the American entrepreneur, and Stavros Niarchos and Aristotle Onassis, the two Greek operators. Next to Ludwig, Onassis and Niarchos are the biggest independents in the business. The Greeks seldom run their ships under the Greek flag, partly because of the vicissitudes of politics and partly because they want to avoid possible taxation at home. Instead their ships are owned by front companies in Liberia, Panama or Honduras where there are little or no taxes. In the tanker business, it is usual for an operator to win a contract to haul oil from one field to a specified market. The charter is used as collateral with a shipyard. Most of the ships are constructed in Japan, Scandinavia or England; as Western European labor costs rise, some of the Greeks have actually begun to build ships in their homeland. In recent years, Niarchos and Onassis have also become involved in combines with oil companies. Increasingly, with tankers as their base of operations, they act as brokers for Japanese and American oil and ship building concerns which have extra cash for investment.

The closing of the Suez Canal during the Arab-Israeli War of 1967 really provided the major impetus for building the new fleet of super tankers. When the canal was closed, and the ships were forced to go around Africa to reach the European markets, shipping rates went sky high. To bring about a reduction the oil companies set about building the enormous ships which could hold enough cargo to bring the unit costs down. As a result the charter rates now actually are less than half what they were when the canal closed in 1967. The Suez Canal now is virtually obsolete because the new ships, with drafts of sixty-five feet or more, can not negotiate the channel. For that matter, they are too deep draft to sail fully laden into most ports of the world. (There are but nine ports which can handle the new ships fully laden: two in Scandinavia; three in West Europe; Bantry Bay, Ireland; the Welsh port of Milford Haven; Finaart on the west coast of Scotland; and Tokyo Bay.)

As the tanker fleet grows, the risk of accidents at sea increases. A group of British experts who surveyed the *Torrey Canyon* disaster reported: "The risk of accident is a very real one. In the three years preceding the wreck of the *Torrey Canyon*, 91 tankers were stranded in various parts of the world, while 238 were involved in collisions either with tankers or other vessels. Over the world at large, tankers thus have been involved in potentially serious accidents on an average of about twice a week for the past three years [prior to 1967]. Sixteen of the 329 ships which were concerned became total losses; in nine of the collisions fires broke out in one or both ships; and in 39 cargoes spillage or leakage occurred."

Oil spills from tankers are a major source of pollution. Max Blumer, senior scientist at Woods Hole Oceanographic Institution, in Massachusetts, described the Institution's experiences in the southern Sargasso Sea in 1968. "During a recent cruise of the R/V Chain of the Woods Hole Oceanographic Institution to the southern Sargasso Sea, many surface 'Neuston' net hauls were made to collect surface marine organisms. . . . Inevitably during each tow, quantities of oil-tar lumps up to three inches in diameter were caught in the nets. After 2–4 hours of towing, the mesh became so encrusted with oil that it was necessary to clean the nets with a strong solvent. On the evening of 5 December, 1968 . . .

the nets were so fouled with oil and tar material that towing had to be discontinued. It was estimated that there was three times as much tar-like material as Sargasso Weed (on a volume basis) in the nets. Similar occurrences have been reported worldwide by observers from this as well as other institutions."

One billion tons of oil is shipped by sea every year, and much of that is hauled through busy stretches of channel. The most heavily traveled passages are in the Gulf of Aden, the Malacca Strait, between Singapore and Indonesia, the English Channel and the Florida straits. (As an example, 25 percent of the total world oil shipped goes through the English Channel.) Based on losses during various marine accidents, Blumer concluded that the oil influx into the ocean is at least one million metric tons a year (shipping losses only) and it is likely to be considerably higher.

It is staggering how much petroleum goes overboard. The *Torrey Canyon,* which broke up in 1967, washing the coasts of England and the continent with oil, is perhaps the best-known recent instance. Here are some other major spills since 1966:

1966: The tanker *Seestern* pumped 1,700 tons of oil into Midway Estuary, Kent, England, killing thousands of birds.

1967: The *Desert Chief* spilled 1,200 barrels in York River, Virginia, killing thousands of water fowl.
—Heavy oil slicks hit thirty miles of Cape Cod, killing birds.

—Seventy-five different incidents of oil pollution were recorded in Cook Inlet, Alaska.

1968: *Ocean Eagle* splits up off Puerto Rico, flooding the coast with one million gallons of oil.

—*World Glory* broke in half off South Africa spilling 46,000 tons of oil onto the coast.

1969: *Hamilton Trader* collided with another vessel, gushing oil all over the north Wales coast.

—The French tanker *Gironde* collided with an Israeli ship, spilling 1,000 tons of oil onto the Brittany coast.

For all of 1969, the US Coast Guard reported there were 1,007 oil spills in US coastal waters. The Federal Water Quality Administration estimates there may be as many as 10,000 oil spills each year. Many oil spills, of course, are never reported, and it is standard practice on ships to pump waste oil into the water at night on approaching port.

During the first three months of 1970, oil spills from tankers and leaking offshore rigs were running at a much higher rate. Again, the most prominent instances:

February 4: The Liberian tanker *Arrow*, owned by Aristotle Onassis under charter to Standard Oil of New Jersey, hit a rock off Nova Scotia, broke her back and sank, spilling some 2 million gallons of heavy fuel oil into the Atlantic.

February 10: An intensive fire, burning out of control, was reported off Louisiana on a Chevron Oil rig. Ten wells were out of control. Fire was eventually snuffed with dynamite; wells spilled oil in a twenty-mile-long spill.

February 10: An intensive fire, burning out of control, was aground in Tampa Bay and started gushing heavy fuel oil from a hole in her hull: 21,000 gallons flowed into the sea and down on a mangrove swamp, killing all forms of wild-life, including 11,000 birds.

February 26: Freighter *Marc Buccaneer*, inbound to Jack-sonville, Florida, slammed into the oil barge *Eastport*, ripping a gash below the barge's waterline, and spilling 7,000 gallons of heavy oil. The spill killed birds and was carried into a wilderness preserve.

March 3: Liberian tanker *Oceanic Grandeur* broke its bottom in Torres Strait off Australia, spilling 58,000 tons of oil.

March 3: Ten thousand dead birds were discovered strewn over 1,000 miles of twisting Alaskan shoreline. They were drenched in oil, which came from a slick of undetermined origin. Federal officials suspected the slick was discharged on the high seas by tankers and blown ashore in a stiff breeze.

April 1: Storage tanks of oil at US Navy's Kodiak Base, Alaska, overflowed spilling 45,000 gallons of oil into Women's Bay, coating the beaches and killing wildlife.

April 26: The Coast Guard spotted two more oil slicks in the Louisiana Gulf. One spill, thirteen miles long, was coming from a leaking barge which served as a collection tank for oil wells. The second spill, six miles long, was coming from two offshore platforms.

Aprirl 26: Spill along fifty to one hundred miles of coast in an isolated area of Alaska, northeast of the Aleutian chain, killed tens of thousands of birds, numerous seals, otters and whales.

April 29: The Coast Guard reported a large slick of un-determined origin and size bearing down on the Pensa-cola, Florida area.

While instances of serious oil pollution from ships ramming one another or going aground are becoming more numerous and more serious, the methods for handling such catastrophes have not advanced much over the years. The most ingenious, if desperate, efforts to cope with oil pollution were undertaken by the British when the *Torrey Canyon* broke up in 1967. That ship, laden with 119,000 tons of oil, steamed at full speed in broad daylight onto the rocks off Land's End, England, March 18. After she grounded, the crew first attempted to float the *Torrey Canyon* by jettisoning oil. After 30,000 tons were pumped overboard, it was evident that this technique would not work. A flotilla of ships took up station around the stranded tanker and began to spray detergents on the oil slick in hopes the detergent would break up the oil. (Spraying detergent is the most commonly used method of treating oil slicks. It seldom works and increases the danger to underwater life because it is so toxic.) A Dutch salvage team arrived, and after making a thorough examination of the situation, it was agreed there was a reasonable chance of salvaging the ship. The idea was to wait for high spring tides, due March 26. Gales hampered the salvage efforts, but on March 26, four tugs moved into position and took in the lines from the tanker. The main cable immediately snapped. Shortly thereafter the *Torrey Canyon* broke her back, and more oil poured out. Salvage efforts were abandoned. Attempts to ignite the slick and get rid of the oil that way were futile. Crude oil is heavy and does not ignite easily. Finally the British Air Force was called in, and after bombing the ship for three days, it succeeded in setting some of the oil afire with napalm.

Oil from the *Torrey Canyon* blackened miles of beaches along the English coast, killing birds, injuring the undersea life and ruining the tourist business. The slick spread across the channel to France and raised havoc there.

After the *Torrey Canyon* there was considerable interest in oil pollution, but nothing much happened. In 1969, a well located on the outer continental shelf went out of control off the Santa Barbara, California, coast. All the efforts made to contain the oil there proved just as futile as they had in dealing with the *Torrey Canyon*. After Santa Barbara, the US developed a special emergency oil-pollution contingency plan, where in the event of a big spill, teams of experts from different parts of the government would rush to the scene to view the damage.

When the Onassis tanker *Arrow* hit Cerberus Rock off Nova Scotia February 4, 1970, the *Torrey Canyon* situation was repeated. Detergents were sprayed on the slicks with no effect. The US Army unsuccessfully sent in teams of flame throwers to set the oil on fire. Chemists from Pittsburgh Corning Glass arrived with sacks of little glass balls intended to act as wicks for burning the oil. They did not ignite. Fiberglass collars set up to keep the spreading oil out of a fish-processing plant failed. Attempts to pull the ship off the rock were futile. Finally, amidst a gale, the *Arrow* broke her back, and her stern slid off the rock and sank in one hundred feet of water. The stern went down filled with one million

gallons of congealed crude oil. In April, a salvage team
succeeded in pumping the oil out.

In Alaska, in February 1970, the Federal officials ob-
served the oil slicks hitting the coast and talked about
new-fashioned kinds of cameras which would tell them
which ships discharged oil overboard. In the Louisiana
Gulf, they again watched as crews snuffed out the fire on
ten wild wells. As the fire went out, the oil poured from
the uncontrolled wells, and made a fast-spreading oil
slick. Feeble dams, made of barges strung together
with plastic and plywood, broke in a gale; when they
were hauled back into position, the oil slicks flowed
through and around them.

Since 1968, Senator Edmund Muskie has sought to
add liability amendments to the water-pollution laws.
These amendments never encountered much difficulty
in the Senate, but until 1970 they could not get past the
House, because of resistance from the oil industry. However, after the Santa Barbara spill, there was renewed
interest in the subject, and in 1969, both the House and
Senate passed different versions of liability amendments,
and worked out a compromise in conference.

In the Senate version of the bill, spilling oil from a
vessel into navigable waters in the contiguous waters of
the United States would be illegal. (The contiguous
zone usually runs from the shore out to twelve miles.)
In such instances the ship's owner would be liable for
damages up to $14 million. To escape liability he would
need to prove the spillage was caused by war, an act of

God or by a third party. The House version of the same legislation looked at the problem, the opposite way around. The House bill said the government must prove the discharge is "willful or negligent" before liability could be established. These may at first seem to be technical details, but in effect the Senate begins to place the burden of proof on the shipping or oil company, where the House leaves the burden of proof on the government.

The oil and shipping companies worked hard to kill the Muskie bill. They lobbied through the American Petroleum Institute, the American Institute of Merchant Shipping, and even went so far as to drag in the London insurance brokers who claimed the insurance market couldn't stand the burden of a $14 million liability per ship. While the Senate passed the liability amendments well before Christmas 1969, the House was slow to act, and when it did, the Public Works subcommittee under John Blatnik dragged its feet in bringing the matter to conference. Blatnik was under pressure from the Republicans, primarily William C. Cramer, of Florida, who has vigorously argued the oil industry's cause. Cramer finally buckled in the conference. The big oil spills occurred in his district and he was persuaded to change his position.

While the Muskie version of the legislation, which became law, is regarded as a stiff measure, it really amounts to modest reform. It does not embody a strict notion of liability. It contains no protection for citizens, com-

munities, businesses or other third parties injured because of an oil spill. In fact, oil spillage remains a handy,
relatively inexpensive way of killing off shore communities, ruining business, running people out, and in
that way opening up the coast for more oil operations.

In addition, the Muskie legislation contains a loop-
hole. While the legislation affects *vessels* which spill oil
in contiguous waters, it does not affect oil rigs located on
the outer continental shelf. The shelf usually runs from
the three-mile limit where state control ends on out to
the edge of the shelf which may be fifty or several
hundred miles. The Interior department administers
the outer continental shelf, and the rigs are covered by
liability regulations proclaimed by the Secretary of Interior, Walter Hickel, in August 1969. They make the
operator liable for cleaning up the oil from spills. However, injury suffered by third parties involved in oil
spills falls under applicable laws, which are state laws.
Most of the offshore wells are drilled off Louisiana and
Texas, and neither of those states have strict liability
laws.

In short, under the different US laws, oil companies
and shipping firms are liable for oil spills only in so far
as they must clean up the oil. They have no real liability
to third parties. Injured parties must bear the burden of
the court cases against the giants in the oil industry.

In recent years a new and deadly menace has appeared
on the high seas in the form of vessels carrying liquified
natural gas.

Natural gas is very much in demand as a fuel for generating electricity in California and along the northeast coast because it is relatively free of pollutants, and often cheaper than other fuels. Normally the gas is transported under pressure through pipelines from the gas fields in the South and Southwest. But pipeline costs are rising steadily, and the oil companies, which produce most of the gas, insist they are running short of the fuel. Gas might be piped in from the big finds in Alaska, but the oil companies find it more profitable to sell that gas to Japan.

As the cost of shipping gas by pipeline has risen, there has been a surge of interest in hauling the fuel by ship from other new gas fields in North Africa or Latin America. In order to do this, gas is frozen and packed into special tankers. East Coast utilities already are buying Algerian natural gas and having it shipped by tanker and there are pending deals to bring in gas from Venezuela.

In 1970, there were 237 ships engaged in the gas trade, with twenty-nine more under construction. The trend is to build ships which carry gas in a liquified form. In liquified state, gas is intermingled with other hydrocarbons, propane and methane. After delivery they can be separated from the natural gas, and sold to industrial markets. These gas ships are equivalent to the 200,000-ton class of oil tankers. The key to making money in gas, as in oil, is to keep the unit cost down through shipping in large-bulk quantities. Mobil Oil predicts the US,

Western Europe and Japan will be importing 8 to 11
billion cubic feet of gas daily by 1990.

Importing gas by super tanker poses some bizarre
pollution problems. A report on spillage of liquified
natural gas in marine transportation was issued by the
Bureau of Mines in February 1970. The Bureau sought
to find out what would happen if large quantities of
liquified natural gas were suddenly dumped into the
water from a spill or collision. There is little existing information on gas spills. To begin with, the Bureau re-
seachers dropped a little liquified natural gas into an
aquarium. These tests provided interesting technical
information. However, in one drop the gas exploded
and blew up the aquarium. In another experiment, the
Bureau filled a bucket with liquified gas and hung it out
over a pond by means of a crane. A man on shore then
pulled a lanyard, upending the bucket. As the gas hit the
water, there was a thumping explosion.

The Bureau of Mines concluded that in most respects
transportation of the liquified gas by ship appeared
feasible. However, the Bureau report adds: "Unfortu-
nately, the study raised questions on one aspect of the
problem for which no answers are yet available. Small
scale explosions occurred when LNG was poured onto
water; no explanation can be offered with confidence for
these explosions, and no assurance can be offered that
these explosions could not scale up to damaging propor-
tions in a massive spill." The Bureau of Mines suggested
it might be worthwhile taking another look into this
situation. Meanwhile the fleet of gas tankers grows; these

floating bombs sail in and out of congested harbors on both Atlantic and Pacific coasts.

Oil and gas supply 75 percent of the energy in the United States; coal, 20 percent; and hydropower, 4 percent. Nuclear energy now provides less than one percent of the energy, and while expectations are that its share will increase, thus far development of nuclear power has been extremely slow. According to many predictions, energy demand will triple in the next twenty years. The fastest growing form of energy is electricity, and it is expected that demands for electric power will double each decade. Half of all electricity produced is used by heavy industry.

The main questions consequently are what fuel or combination of fuels will supply the energy, how will this be done, and who will make the decisions. Environmental pollution created in the production of energy turns on this central series of policies.

The myth is that the five major energy sources—coal, gas, oil, uranium and water power—will compete with one another in a regulated market place to provide efficient, safe energy sources at the lowest prices.

Actually, competition in the energy markets is a myth. Even that myth is fast eroding as a handful of large trusts move in to finish cornering the market.

The petroleum industry, as already indicated, is dominated by a handful of large companies which produce

crude oil, shunt it through pipelines, refine it, and dis-
tribute the different end products down through the re-
tail level. The same companies which dominate the
world markets dominate the US markets.

The top oil companies produce most of the natural
gas. Historically, the search for oil and gas are insepa-
rable. Recently, the petroleum companies have spread
their interests into other parts of the energy markets. Of
the twenty-five largest oil companies, eighteen are involved in uranium mining or research. Eigheen of them
have interests in oil shale, a future source of oil. In the
past few years, oil companies purchased coal companies.
Eleven of the top twenty-five oil companies own coal
firms, and virtually all the major oil companies are in
the business of producing chemicals.

But the petroleum industry's domination of fuels is
even more pervasive than these figures suggest. Thus, in
uranium, Kerr-McGee controls 23 percent of the entire
uranium milling business. Humble, Jersey Standard's
subsidiary, is planning to build a uranium mill and
Atlantic Richfield has entered the milling business.

In coal, Humble and Kerr-McGee are building up
substantial coal reserves. Humble purchased large portions of the remaining reserves in Illinois, and is providing coal to utilities in Chicago. Humble and Consolidation Coal, Continental Oil's subsidiary, are
believed to be the only two coal operators east of the
Mississippi with large enough reserves to be able to
supply big utilities complexes, that is, provide enough

coal to fire electric power plants turning out 4 million kilowatts each.

The move into coal began in 1965 when Continental Oil Company, the eighth largest oil producer, announced it was merging with Consolidation Coal Company, the second-largest coal producer, for $600 million.

This merger illustrates how the old trusts have reorganized and are pushing out as energy combines. Consolidation Coal was an amalgam of three venerable Republican baronies: The Hanna Mining establishment led by George M. Humphrey, Eisenhower's Treasury Secretary, and the Loves and Mellons of Pittsburgh. Humphrey engineered the growth of Hanna Mining from a small mining and shipping company into a large industrial organization. In 1929, Humphrey merged the Hanna iron ore and coal properties into National Steel; in the same year he began to organize a large coal enterprise. First, he joined Hanna mines with those owned by George Love. Next, Humphrey picked up large amounts of old Consolidation Coal Company stock, then held by the trust department of National City Bank. After that Humphrey stitched together the Hanna, Love and Consolidation holdings into one big company which kept the name Consolidation. Love became an active officer of the new Consolidation, and he completed development of the enterprise by merging it with the Mellon-owned Pittsburgh Coal Company, then the second largest coal producer. The Mellons thereupon came on the board of Consolidation. Finally, Love made peace with the warring United Mine Workers Union. He made the deal between the coal industry and John L. Lewis in the early 1950s which opened the way for automation in the mines, and made possible the intertwining of union and industry finances and politics through the National Coal Policy Conference, which he created and dominated. The union henceforth served as an ally of the big coal companies in their efforts to monopolize the coal fields.

Love revitalized the depressed coal industry by mechanizing mines on an unprecedented basis, and introducing the strip shovel, which leveled whole mountains at less cost than hauling coal from underground. In doing so, Love demonstrated that coal mining was again a profitable investment for large companies with enough capital to automate. But, in the process, he turned acres of beautiful country into heaps of dirt, impoverishing thousands of workers by putting them out of jobs. However, the profits from Consolidation's mechanized operations enabled Love to acquire a controlling interest in Chrysler Corporation.

Consolidation's 1965 merger agreement with Continental Oil called for exchange of 2 mllion Continental shares to be distributed to Consol stockholders, that is, to be distributed among the members of the Humphrey-Love inside group which held the mapority of Consolidation stock. That block of stock amounted to about 5 percent of Continental stock, enough of a percentage to assure the Humphrey group of a significant voice in decision making.

The reasons for the merger were not hard to find. In

announcing the deal, Continental said the oil company was interested in the "attractive long-term prospects" for coal, particularly for electric generation. Consolidation Coal had recently bought extensive holdings in the coal fields of the mountain states. The West long has relied heavily on hydroelectric projects producing cheap electricity, but there are few damsites left; cheap nuclear power still looks a long way off, and consequently both coal and natural gas stand to be used in increasing amounts in the production of electricity. At the moment they compete with one another. Boilers at generating stations can be switched from one to another depending on price. Continental is the ninth-largest natural-gas producer in the country, and with its hands on new coal fields in the West, it is in a position to influence development of both fields.

Company sources gave another reason for the merger: Consolidation Coal had an important edge over competitors in coal-to-gas research. The Interior department had given Consolidation $10 million in research funds to design, build and operate a pilot plant to make gas from coal at Cresap, West Virginia. The Bureau of Mines claimed gas could be made from coal at costs ranging from 10½ cents to 13 cents a gallon. It cost from 12½ cents to 14 cents to refine crude oil into gas.

(Making coal from gas has a lengthy history. In 1926, I. G. Farbenindustrie, the German chemical combine, announced it had found a way to make gasoline from coal. The Germans had plenty of coal but little oil. This announcement upset Standard Oil of New Jersey, which feared the Germans actually might become self-sufficient in gasoline production and the two giants made a deal promising, in effect, not to meddle in the markets of one another. Joint companies were set up for research and patents were exchanged. This led to the Germans getting the technical know-how from Jersey Standard that enabled their air force during the Second World War to depend almost entirely on gas made from coal.)

The oil companies want to control coal partly because coal competes with oil and natural gas, but also because they recognize the potential for turning coal into gasoline. Gasoline is the major money maker for the oil men, and they want to keep tight hold of any development in that field. To do so, oil firms buy coal companies; they also lobby hard within the Federal government to make sure development of coal is forestalled.

Development of alternate energy sources depends a good deal on how much research is carried out. While the government spends billions of dollars each year on weapons and other war-related research, it devotes only $368 million a year in research into conversion and transmission of energy. Of that total, $311 million goes to developing nuclear power, but only $26 million for investigation of liquifaction of coal. In the past most of that money went to two coal companies—Consolidation and Pittsburgh Midway—which are owned by petroleum combines. The effect of this is to stall off development of coal as a fuel.

The public domain contains vast coal lands. The coal-rich lands are concentrated in six western states—Oklahoma, New Mexico, Arizona, Utah, Wyoming and Colorado.

The government might use these lands to influence the development and transmission of energy, according to a public policy. Instead, under both Johnson and Nixon, the government virtually gives away the land to the big oil companies, which squat on it without working the coal. This extraordinary operation is directed through the Interior department's Bureau of Land Management.

The Bureau of Land Management has charge of some 500 million acres of public land, and it administers the land in accordance with different laws. Under the mineral leasing act of 1920, anyone may prospect for coal on Federal lands in blocks of up to 5,000 acres for a total of 46,000 acres in any one state. The prospector applies at the nearest Bureau of Land Management regional office, pays $10 for each 5,000-acre block and 25 cents an acre rent for one year. He thereupon controls the minerals in that land for one year. The permit can be renewed for a second year on the same terms. If the prospector can persuade a geologist and mining engineer to say the find is workable, he can obtain a preference-rights lease to the land. The mineral rights are his forever, so long as he pays a rent of 25 cents an acre for the first year; 50 cents an acre for the second, third, fourth and fifth year; and one dollar an acre for the sixth year.

If the prospector actually mines coal, he need only pay the government a royalty of 17 cents a ton.

These regulations make it a cinch for any modest business endeavor to capture and sit on valuable coal lands for at least eight years for an insignificant amount of money.

The results of the system are clear enough: only 10 percent of the coal leases granted by the government are ever worked. Those leases are held, to a large extent, by the huge petroleum combines. Included among them are Kerr-McGee, Continental Oil, Gulf Oil, Atlantic Richfield, Mobil and Sun Oil.

In 1969, shortly after the Nixon Administration took over, a group of economists within the Bureau of Land Management made an attempt to jack up the price of Federal coal lands. They wanted to find a method which could reduce the amount of leasing, and at the same time ensure the land was actually worked and that the government got a proper price for it.

Where two or more parties want to mine coal on a certain lot of land, the Interior department sets up a bidding. Through the years, Congress has insisted and the Interior department has dutifully put into practice an old-fashioned oral bidding formula. Sealed bids are submitted, but they are not important: the government never establishes an "upset price," that is, a price below which bids will not be considered. The bidding is run like a poker game. The competitors usually send in very low sealed bids. The day of the land auction, they come along to the Bureau offices and sit around a table with the auctioneer. The sealed bids are opened, and the competitors watch the government's face for a reaction. They then begin to cautiously bid up a bit from the low sealed bids. The bidding goes around the table, and stops at the highest bid.

In 1969, in Wyoming, the government required sealed bids from everyone and established an "upset price." The result was that Energy Resources, a subsidiary of Iowa Power & Light Company, bid $35 an acre for coal (about two thirds more than Interior usually gets for coal lands), and Consolidation Coal, now Continental's subsidiary, bid only $5. Interior economists were pleased with their success, and they scheduled a second bidding for coal lands in Montana in February. Unexpectedly, lawyers from one of the interested parties in this auction showed up in Washington. They represented Peabody Coal, a subsidiary of Kennecott Copper, and they protested at the idea of sealed bids. The lawyers pointed to language in the law which they claimed prohibited the practice of sealed bids. Thereupon, Assistant Secretary of the Interior for public lands Harrison Locsch canceled the Montana sale, and removed the coal-leasing program from the Bureau. Instead, he gave control of the operation to the US Geological Survey, which always has been more prone to take industry's point of view. In the past, the Geological Survey argued for the lower royalty rates.

Loesch, a former attorney from Montrose, Wyoming, believes the Bureau of Land Management had gone too far in demanding such high prices from industry. Nixon

picked Loesch for the job on the advice of Gordon Al-
lott, ranking Republican on the Senate Interior Committee. Loesch takes what is known in Washington as
the "more flexible approach" on public lands. He explains his views this way: "Mining is essential to this
country, and it will be continued on public lands. We
aren't putting any new restrictions on miners, but we expect them to respect the land values and to rehabilitate
the land when they're through with it."

Next to oil, natural gas is the largest source of energy
in the country. It accounts for about 32 percent of all
energy. It supplies 23 percent of the electricity, the fastest growing energy market.

In 1954, the Supreme Court directed the Federal
Power Commission to regulate the price of gas at the
well head. During the 1950s, the Commission sought to
avoid the court's order, and worked hard to sponsor
legislation exempting gas producers from its control.
The legislation failed and the court repeatedly ordered
the Commission to act. Finally, in 1962, the FPC set out
to create a pricing system by setting a price for a given
area where gas was produced. Now, it appears on the
verge of abandoning the system.

Shortly after he became chairman of the Federal
Power Commission in the summer of 1969, John N. Nassikas said he thought the price of natural gas should be
increased. Nassikas claimed the nation's proved reserves
of gas were not increasing at a fast enough rate to meet

the demand, and that, as a result, the supply of gas would be exhausted in ten years.

Nassikas based his estimates on a commission report which in turn based its conclusions on industry data. Neither the FPC nor the Interior department, which also administers gas policy, make their own independent studies. In this case, the FPC never saw the actual confidential company data. Instead, that information was screened and summarized for the commission by the American Gas Association, a trade group. The FPC report says openly: "For the purposes of this report, we have accepted at face value all industry-furnished supply data." The Interior department reached a similar conclusion after a one-day meeting between officials and gas-company executives.

The reserve statistics of the industry are misleading since they do not take into account the huge finds of gas on the north slope of Alaska, nor do they account for large finds on the outer continental shelf off Louisiana. Nor do they reflect estimates of the US Geological Survey, which indicate gas reserves in the United States total 3,655 trillion cubic feet. At the present time, total production runs about 20 cubic feet annually.

Of more immediate interest, however, is the fact that there are 500 shut-in gas reservoirs with drilled wells on Federal leases on the outer continental shelf off Louisiana. A shut-in reservoir is one in which a well has been drilled and completed, that is, it is capable of production—but for one reason or another is not producing. That might be because of pipeline connection or because of production difficulties. If these zones were connected to consumer markets, it is estimated they could supply gas for household-heating purposes to 4 million homes. That is a considerable number considering that all such homes now served by gas total 29 million.

While the Federal leases on the outer continental shelf require that minerals be developed in a proper and "timely" manner, the US Geological Survey, the agency within the Interior department which administers the leases, pays little attention to the producers and leaves them alone. The lease lasts five years, and right now producers are drilling well in 3½ years. If there were indeed a serious gas shortage, that amount of time would hardly qualify as a "timely" drilling program. As the petroleum companies play the gas game, the object is to find gas, then keep the lease while at the same time holding the gas off the market, hoping a short supply will drive the prices up. This is permissible under the government regulations, but it is a little tricky. What happens is this: a company discovers gas on one of its leases. It drills a well and starts production. Then, allegedly for the sake of efficiency, it pulls all its other leases in the area together and creates a "unit" around the one producing well. Normally, it would be required to drill a well in each lease area within five years. But once there is a unit, it is only necessary to have one well for the whole unit. In this way huge supplies of natural gas are held off the market.

Speculation in natural gas is not a game played by small operators. To operate on the outer continental shelf, companies must pay millions of dollars in so-called "bonus bids." Little operators are eliminated, and only the big firms can play.

The Nassikas line, that by providing a price incentive to industry it will work all the harder to find more gas, is ludicrous. Since the search for gas is carried on by huge petroleum conglomerates, usually in connection with a search for oil, there is no assurance whatever that added revenues from gas would go to bolster a further search for gas. In all likelihood it would go into whichever area of corporate activity that would be most profitable. That might mean the development of a new kind of paint or cosmetic, and have nothing to do with gas.

So regulation, with the Federal Power Commission running the industry's Washington publicity office, becomes a method of manipulating the energy markets. There are other ways in which the energy markets are manipulated to maintain pollution. Nearly half of all crude oil is used for gasoline, which, of course, explains why both the oil companies as well as the auto industry are so attached to the internal combustion engine. The automobile accounts for 60 percent of all air pollution, and in congested urban centers, autos contribute as much as 85 percent. Nearly all carbon monoxide comes from vehicle exhausts. Each year autos produce 90 million tons of pollutants, twice as much as any other source.

There are two obvious methods for dealing with air pollution. One is to build rapid-transit systems in large cities to compete with and eventually replace automobiles and trucks. The other is to develop an alternative type of engine which does not cause so much pollution. There has been little interest in constructing mass-transit systems on any sort of wide-scale basis, mainly because of pressure from the automobile industry and other interests concerned with building and maintaining highways.

Whatever type of alternative engine is developed, it would probably use fossil fuels. Even so it is possible to make engines which vastly reduce pollution. In this sort of situation, the usual approach is for the government to sponsor research, then contract with a manufacturer either for a prototype, or for a fleet of test vehicles. That is the way armaments are designed and built. In this case, the government can actually create a market since it buys 60,000 vehicles a year.

But the auto and petroleum industries do not want to change the type of engine, and neither the Congress nor the Administrations in the last eight years have been seriously willing to challenge them. In the early 1960s, the auto makers claimed vehicles made but a slight contribution to air pollution. Then they announced research programs, and now are claiming they will clean up all pollution from vehicles by designing more efficient internal-combustion engines and adding emission control systems. That program, now underway, has resulted in equipping autos with exhaust-control systems which the Chrysler Corporation, for one, had developed and installed on a fleet of cars in 1962. The emission standards, set by the government, only pertain to new cars, which are a fraction of the total automobiles in operation. Thus, installation of emission-control systems may slow the growth of air pollution, but it won't stop or reverse the growth. Moreover, installation of emission controls will cost an extraordinary amount in maintenance. Some estimates run as high as $6 billion a year.

Carbon monoxide, oxides of nitrogen, leads and oxidants are the primary components of vehicle air pollution. The government first set emission standards for January 1968. Progressively stiffer standards have been proposed for 1973 and 1975. According to a report by the Senate Commerce Committee: "The present emission standards will not stabilize, much less reduce, vehicular air pollution. Studies indicate that, under existing controls, automobile air pollution in the United States will more than double in the next 30 years because of the projected increase in both the number of vehicles and miles driven by such vehicles. Ironically under the present emission standards, oxides of nitrogen emissions, the main villains in the photochemical smog production, will be higher than they would be if no standards existed." That is because current standards regulate hydrocarbons and carbon monoxide only. The best way to reduce these pollutants is to raise the combustion temperature, which, in turn, increases oxides of nitrogen production. Even if devices were effective, there could be no assurances that they would be maintained through the life of the car. Beginning in 1973, the government intends to establish standards for oxides of nitrogen. There is no indication from the auto makers as to whether or not they can or will abide by the standard.

In 1967, the Senate Commerce Committee began an inquiry into the feasibility of alternative engine systems. It began with a look at electric automobiles. They did not seem a sensible alternative because the batteries were too bulky and expensive. Eventually the comittee and a special study group in the Commerce Department recommended development of a steam car, driven by a Rankine-type engine. In a Rankine engine, fuel is ignited outside the engine in a burner. The heat produced by burning heats a fluid—water or some other neutral fluid—and is thereby converted to a vapor. The vapor becomes the power source for the engine. Valves are opened and shut to control entry of the fuel into a shaft where it turns the pistons which drive the wheels. The Rankine-type engineer car runs on a variety of fuels. It can burn gasoline from a pump or it can operate just as well on low-grade white kerosene. Rankine fuel is less refined, less costly and less polluting than fuels which go into an internal-combustion engine. So far the one real drawback to the steam system is its high cost.

Initially, there was considerable enthusiasm for the steam engine, particularly on the part of William Lear, the man who invented the Lear jet. Congress became fascinated with the possibility that this independent inventor might be persuaded to turn his genius to making a new kind of engine. Lear plunged into development of a steam engine. He claimed to have spent about $6 million in developing a prototype steam car. The idea was to enter the steam auto in the Indianapolis 500 where it could make a big splash by whipping all the other entries. However, by late 1969, Lear still had not unveiled his prototype steam car. In November, he told *The New York Times,* "I've thrown in the sponge on steam." He claimed there was only "one chance in 500,000" that a replacement could be found for the internal-combustion engine, and "I don't think it can be found." He went on to say, "I've been billed as the great champion of the steam car, and I've got 5,500,000 reasons why I'm not." "No matter what you do (with steam) you'll never come up with something that is good for the public." And he concluded the idea of a steam car is "ridiculous—no one's going to do it."

Lear later told the Senate Commerce Committee that the newspapers had mistaken his position. He had come to believe it was useless to think of competing with the automobile companies in any substantive way. They would never purchase engines or parts of engines from outside companies. Therefore, he thought the most sensible solution was to encourage the auto industry along research lines it already has begun. Instead of steam, Lear now thinks there is a possibility in the gas turbine, an engine which is relatively low in pollutants, and which the auto industry is busy developing.

The steam-car charade went according to American myth. First there was the dispassionate, scholarly, objective research which found it was all possible. Then there was the tough-minded, hard driving Horatio Alger inventor in his Nevada shop. The end result has been to give the auto makers an even firmer hold on the monopoly for internal-combustion engines. Meanwhile, pollution is worse than ever.

Instead of a serious attack on air pollution, the government is pursuing a small, meaningless program. In 1970, Congress appropriated $108 million for the National Air Pollution Control Administration, part of HEW. Of that amount, $38 million went for research and development. Only $9 million of that was committed to research on vehicles. In the past few years the government has spent no more than $500,000 a year for research on alternative engine systems. In 1971, Nixon has proposed a reduction in research funds from $38 million to $27 million. Thus, while autos are responsible for the great bulk of air pollution, a tiny proportion of the budget for pollution programs goes for work on automobiles.

Recently, a group of senators, led by Magnuson, sponsored legislation to establish a procurement scheme, whereby the government would create a fund of $50 million in order to purchase vehicles with an alternate engine low in pollutants. In effect, the proposal would create a test market for industry. The manufacturer could sell vehicles at 25 percent more than usual. Now, the hope is that Lear will be successful in making a gas-turbine engine, which will find acceptance at American Motors. There the management is looking for a means of hauling the firm out of the dumps.

Thus, the committees of Congress act out a façade of

reform which so far has assured automobile makers continued use of the engine and fuel which cause all the pollution.

THE RUIN OF SANTA BARBARA

While the giants in the petroleum business also control coal, uranium and natural gas, they drill for oil because they can make more money that way. The biggest profits come from selling refined petroleum products, especially gasoline. The government encourages the oil companies in pursuing this course through an elaborate set of economic incentives including tax write-offs and import quotas. Because of these economic incentives, the oil men have been in a frenzied search for oil that in the last five years has led them to underwater explorations of the continental shelf areas surrounding the continents of the world.

Since 1954, the US has claimed control of the outer continental shelf (OCS) extending offshore around the United States. In that year the Congress enacted a law reserving the territory for the United States. (In the preceding year it had given away the tidewater oil lands to the states for exploration by the oil companies.) While US authority over the OCS has been challenged by different states, nothing has come of their claims. However, suits are still pending in the courts.

The states control the tidewater which runs from shore's edge generally out to the three-mile limit. The government, through the Interior department, then ad-

ministers the outer continental shelf from the three-mile
limit on out to the point where the shelf turns into a
slope and drops into the ocean depths. That may be fifteen miles, or, along some parts of the coast, several hundred miles from shore.

The more the incentive to drill for oil, the more the
petroleum companies are encouraged to develop and sell
products utilizing oil. The more end products they sell,
the more the demand for oil. But even if one were to
accept as practicable the uses to which oil is now put
(i.e. the manufacture of widening ranges of plastics,
chemicals and foods), there still is no need to drill for
oil on the outer continental shelf. Oil is plentiful in
Canada, Venezuela and the Middle East, and if worse
comes to worse, there is oil in Alaska. There's enough
oil in the Rocky Mountain oil shale to meet US demands
for 400 years. But the import quotas keep foreign oil
out, and the tax incentives encourage the oil men to drill
within the United States. While the Interior department
is hardly aggressive in administering the outer continental shelf, it is anxious to rake in whatever royalty moneys
the petroleum companies are willing to give for producing on Federal lands. Consequently, the government and
the oil industry work together to promote oil exploration on the shelf. So far, Interior has leased out about
one percent of the shelf. (The total shelf area encom-
passes 1.2 million miles.) Most of the leases are off
Texas and Louisiana, along with a small portion off
Santa Barbara where the big spill took place in 1968.

The Federal government exercises very little control

over the leasing. Oil companies first ask for, and are routinely granted exploration permits by the US Geological Survey, the agency of Interior which has charge of technical and scientific aspects of the drillings, and which supervises the actual production. If a company discovers oil and wants to begin production, it asks Interior to place the desired land up for lease. The department usually complies with the request. The Geological Survey does not investigate the proposed lease area, but instead depends on information submitted to it by the company. Since it has not developed any information on its own, the government doesn't know how much oil or gas is likely to be in the area, and is not in much of a position to judge the bids. As a result, the companies pay pretty much what they want to pay. Often times there are no competitive bids (oil leases have been granted on a non-competitive basis in 200 square miles of ocean). Citizens who wish to inquire into the possible dangers from drilling operations by examining the Geological Survey data are prevented from doing so. The information, provided the government by the companies, is judged by the government to be proprietary and hence held in secret. (While the Freedom of Information Act, passed by the Congress in 1968, supposedly lays bare government records for the inquiring citizen, it specifically excludes data held by the US Geological Survey.) During Stewart Udall's tenure at the Interior department, the Solicitor made a move to force the US Geological Survey to make its own seismic and geological invesigations of proposed lease areas, on the grounds that the government ought to know as much about the areas as possible to ensure safety, and to get the best price possible. The Geological Survey resisted the move, and it was eventually killed by the Assistant Secretary for Mineral Resources, Cordell Moore. A geological investigation of an oil-drilling area might seem to be a straightforward proposition, but it was made clear to Udall and the Solicitor from two key points in Washington that this sort of study would not do. The Budget Bureau stated that funds for such a drilling program would be cut out of the Interior budget. And word came from the House Appropriations Committee, headed by George Mahon of Texas, that if Interior pursued this idea, Udall would face a keel-hauling.

The Interior department always has been the lobby for the oil and gas men. The Department functions as the government's real-estate office, and naturally is pleased to see the public domain leased or sold off for profitable use. In addition, the geologists who work at the US Geological Survey within the Interior department are prone to accept the industry point of view.

During the Johnson Administration, the government took an increased interest in the real estate on the outer continental shelf. During 1966 and 1967, Johnson vigorously pursued a juggling act of guns-and-butter program. With one hand, he claimed to be pumping more money than ever before into making America a better place to live, and with the other, he was pouring dollars into the Vietnam war. One of the gimmicks the President employed to lend legitimacy to his juggling was to assure the citizenry that the budget was balanced, and indeed, to show there was even a surplus. He was able to do this in large part because of careful timing of lease sales of the outer continental shelf. The sales produced millions of dollars which briefly created the illusion of a balanced budget. From the point of view of the White House, one of the most important sales was to involve the channel off Santa Barbara, California.

The Santa Barbara channel is different from other offshore oil fields in that it consists of several thousand feet of young unconsolidated porous and permeable sands, which are deposited directly on the ocean floor. Many parts of the channel lack any sort of "cap rock" structure to trap the oil, and the only barrier between the oil and the ocean floor is a layer, no more than one-hundred feet thick, of fine sand and silt. The coating over the ocean floor is thick near shore, but thins as it spreads outward. In fact, at some locations the sandy layers between the sea floor and the oil are so narrow that a ship's anchor can cause the oil to begin to seep out of the ocean floor. The pressure of oil rigs alone, sinking their foundations into the sand, have created seepage. The area is all the more hazardous because Santa Barbara lies at the center of criss-crossed faults which cut deep into the earth's center. Inshore, parallel to the channel, the deep San Andreas fault runs 600 miles up the coast. There are several faults running through the channel islands, and the major San Obispo fault terminates at Santa Barbara. The faults act as passageways for oil, which trickles upward from the oil rocks below into the porous sand on the ocean floor. Santa Barbara is the center for most serious earthquakes in the United States. Three major damaging quakes have been recorded since 1925. In the summer of 1968, before the drilling for oil had begun in the channel, a "swarm" of sixty-six minor earthquakes occurred during a six-week time space.

Geologists and residents of Santa Barbara have argued that the combination of faults, earthquake history and peculiar bottom strata make drilling for oil around Santa Barbara a hazardous undertaking. And they have objected to the petroleum industry settling into the town, bringing on the danger of pollution, and as a practical matter, damaging the tourist business.

The first tidelands oil was refined in Santa Barbara in the 1890s. In 1938, California authorized oil drilling in tidelands out to the three-mile limit, and after the Second World War, several big oil companies made geological studies of the channel floor in anticipation of drilling. Because of the outcry by townspeople, angered at the prospect of seeing Santa Barbara turned into an oil town, the state agreed to establish an offshore sanctuary sixteen miles along the coast and three miles out into the ocean. Leasing in the sanctuary would only be permitted if the state-lands commission found that oil was being drained from outside the three-mile limit. Between 1945 and 1955, the state leased considerable portions of the tidewater. During this time, both the companies and the state learned a good deal about the bottom conditions. (Because of state law prohibiting disclosure, California never shared this information with the Federal government.) Then, in 1965, the Supreme Court rejected California's claim to waters beyond the three-mile limit, and immediately the Interior department announced its intention of leasing the channel. Federal leasing procedures were hastened when the state granted leases just within the three-mile limit in 1966. During that year the then Secretary of Interior Stewart Udall authorized the lease of one Federal tract to prevent any possible oil drainage from the adjacent state platforms.

Various citizens' groups as well as the elected officials of Santa Barbara became increasingly upset as the Federal leasing program swung into action, and during the spring of 1967, they met with Interior officials in an attempt to ensure that drilling on Federal lands would not excite the state into leasing the sanctuary.

Meanwhile, the oil companies spent a reputed $150 million exploring the channel. And Interior, under pressure from local citizens, agreed to extend the sanctuary by adding a two-mile buffer zone in Federal waters. In November 1967, the Corps of Engineers, which has charge of construction in harbors, held a one-day hearing on an application by Phillips Petroleum for a drilling platform. At the time, townspeople tried to argue the dangers from earthquakes and faults, but hearing officer,

Norman Peherson summarily insisted these considerations were not the business of the Corps.

Udall himself was badgered by the Western Oil & Gas Association, a producer group. It claimed the companies had sunk huge sums in development of the channel, and it was not fair to keep them waiting for the oil. For his technical advice, Udall depended on William T. Pecora, director of the US Geological Survey. According to Udall's associates of the time, the Secretary received repeated assurances from Pecora about the leasing program. And whether that is true or not, the Geological Survey stood behind the lease program from the outset.

Most of all, Udall was getting heat from the President through the Bureau of the Budget. In July 1967, while Interior department officials were "negotiating" with Santa Barbara groups about the leasing program, Phillip S. Hughes, Deputy Director of the Budget, told the then Undersecretary of Interior Charles Luce, "As you know, we are particularly interested in the Presidential decision to generate additional revenues from the leasing of the nation's mineral resources. We are convinced that improved efficiency of federal OCS policy, procedures and regulations would lead to lower capital and operating costs for leases. These lower costs would, we believe, be reflected in increased bonuses from OCS lease sales. Recent indications from oil economists, oil companies and in oil trade journals are that economic benefits through improved, more efficient practice would be exceedingly large."

In his reply to this, Luce, who has since become president of Consolidated Edison, pointed out how leasing the OCS could effect the oil-import quotas: "It should be emphasized that increased production on OCS would proportionately increase our revenues only if it did not result in lower domestic oil prices. Thus, to realize proportionately increased revenues it would be necessary to restrict oil imports in approximately the same amount as the increased production, or alternatively, the states would have to reduce their allowable production in that amount." Luce added, "If we should decide not to offer any federal lands for leasing off Santa Barbara the decision would be very expensive in terms of immediate revenue to the United States." (At the time Udall and his immediate staff at Interior were fighting for abandonment of the import-quota program: here Luce is pointing out that the Budget Bureau is binding Interior more tightly than ever to the import-quota system by making revenues dependent on domestic production of oil.)

A few months later Udall received a blunt note from Charles Schultz, director of the Budget Bureau: "As you know, the president has on several recent occasions instructed us to make every attempt to produce additional revenues from Federal resources. Last fall after a successful sale off the Gulf Coast, he specifically asked me what might be done to increase revenue from offshore leasing."

The citizens of Santa Barbara had no luck in persuading Interior to postpone the leasing until detailed studies of the dangers involved could be made. And they had no success in persuading the Department to widen the buffer zone, reduce the number of leases, and restrict the number of drilling platforms. Interior would not budge. Assistant Secretary Moore put it this way, "Further action of this sort would be beyond our original intention to prevent drainage of the state sanctuary. The two affected tracts cover the remainder of the Rincom trend and is the last of the prime acreage of our proposed sale. This would be 'deferring' additional bonus money above that that could be received if we leased the two-mile buffer zone acreage. I cannot see how these people can expect us to absorb any additional revenue losses merely because of their further obsession to protect their view.

". . . At a time when the President has asked that we collect all revenues possible and when we can easily make additional multimillion dollar collections, I don't see how this additional 'deferrment' can be justified to either the White House or the Budget Bureau, and especially in view of the reason that the additional platforms are merely restrictions to the view. I especially hope that the Secretary is not placed in the position of explaining such additional action should the Budget Bureau chance to look into this matter, in conjunction with their recent letter on OCS leasing revenues. I do think that we can defend the sanctuary's protective measure but not any additional revenue deferrals."

Faced with this sort of pressure to produce short-term cash flows for the White House, Udall went ahead and in February 1968 held a lease sale on the Santa Barbara channel, accepting $603 million for leases on seventy-one tracts totaling 363,000 acres.

By January 1969, the US Geological Survey had approved five development wells on Platform A, which was operated by Union Oil. For each of these wells, Union obtained from the Geological Survey regional supervisor a variance on the usual well-casing requirements. The company obtained permission to set one string of conductor casings to only fifteen feet below the ocean bottom, as opposed to the usual Survey requirement of at least 500 feet. More important, a smaller surface casing went to only 238 feet below the floor as opposed to the 861-foot minimum requirement of the Survey. On January 28, with three wells already drilled and shut in, the Union crew dug a fourth well—A–21. The drill crossed two large faults and other smaller faults on the way down and eventually hit a major oil deposit at 3,479 feet. As the crew pulled the drill out, it stuck for a moment, then came free. Suddenly gas exploded up the well to the surface, and for the next thirteen minutes the crew was engulfed in a blinding gaseous mist amidst a deafening roar. Working fast, they "stabbed" the drill back into the hole, shut the blow-out prevention valves at the well surface, apparently bringing the well under control. But with surface escape shut off, pressure built up down the well below the point where the surface casing stopped; as pressure increased, the oil burst sideways out of the well into layers of bedrock. From there it pushed up, buckling the ocean floor, and began to seep out onto the sea bed. The well was out of control beneath the surface of the ocean.

The next day the Coast Guard took charge. Union Oil assumed financial responsibility for the accident. Chemical dispersants, plastic booms, log-type booms, cement and other chemicals were all tried to break up the growing slick. All were futile. By February 4 and 5, heavy crude oil covered most of the Santa Barbara beaches. Straw was ordered to be spread over the oil. On February 3, Secretary Hickel requested that all drilling be halted in the channel, but on February 4 he withdrew the request. Finally on February 7, drilling was halted. Seepage continued through the ocean floor. Within two months of the spill more than one hundred miles of beach were hit by oil in varying degrees of severity. One expert estimated that by the middle of May 1969, about 3,250,000 gallons had erupted.

On February 17, Hickel announced new regulations which provided for "absolute liability." They were subsequently modified and the absolute liability was toned down. A presidential panel on oil spills was convened, and it recommended perforating other platform-A wells in order to conduct temporary pumping to relieve pressure. Hickel asked the companies for more information concerning the channel, promising them it would be held in secret. A second oil-spills panel was convened, and in April Hickel authorized renewal of drilling on five leases in the channel. The second panel recommended resumption of drilling as the best way of alleviating seepage through the ocean floor. Hickel thereupon reviewed the leases, and allowed them to resume drilling. The Corps of Engineers subsequently received requests for drilling permits. In August, the county of Santa Barbara objected to the permits and asked for public hearings. In September, the Corps issued the permits without hearings. The county filed suit in Federal district court seeking a temporary injunction on issuance of further permits pending hearings. The court held against the county and the Corps issued more permits. The county subsequently lost in Appeals Court and was denied a hearing by the Supreme Court. A new platform was placed just east of platform A. It belonged to Sun Oil. In December a new spill occurred in the channel. The pipeline leading from the well to shore began leaking. When the well was shut for repairs, pressure built up, and oil again began coming out through the channel floor. The pipe was finally repaired, and while the seepage was reduced, it nonetheless continued. By that time, Christmas 1969, twenty-two miles of Santa Barbara and Ventura beaches were newly polluted.

Despite the occurrences at Santa Barbara, William Pecora, the director of the US Geological Survey, has steadfastly maintained there is nothing to worry about in drilling across faults in the midst of earthquake territory. In late February 1969, Senator Muskie's subcommittee on air and water pollution made an investigation of the circumstances surrounding Santa Barbara. In the process of doing so, Muskie and Pecora became involved in the following exchange:

MUSKIE: In your judgment, does the history of earthquakes in an area enlarge the possibility of such risks?

PECORA: The history of earthquakes on the West Coast is fairly well known. Earthquakes, in my judgment, sir, are not considered an undue risk insofar as drilling is concerned.

The danger of earthquakes would more reasonably be applied toward the development of tidal waves or tsunamis, as they are called, which can put massive waves upon the shore endangering all installations on shore.

MUSKIE: I am thinking not of the risks from the earthquakes directly but what earthquakes do to the geological formation of the earth. Does the fact of earthquakes in this area raise the possibility of a change in the geological formation of the earth that would enlarge the risk of oil seepages or explosions from core drilling operations?

PECORA: In my judgment, the relationship of earthquakes has been unduly stressed as a danger. California is earthquake country. The purpose of the adjustments that the crust of the earth is going through in the development of an earthquake is a momentary unlocking of a locked position for displacement and during that displacement period the earthquake occurs, the tremors occur.

But the faults are in locked position all through the west coast. An earthquake fault is active only for that period in which that locked position is released and then relocked.

MUSKIE: Do you conclude, therefore, that the Santa Barbara incident has no connections with the earthquake history in this area?

PECORA: I shall state that in my judgment the earthquake history of the area has no bearing on the drilling problem insofar as the risk is involved as per your question. . . . Whether or not there are earthquakes in the Santa Barbara region or elsewhere in California should have no implication as an undue risk factor in drilling.

In April 1969, Pauley Petroleum, Inc., a leaseholder in the channel, sued the Federal government, claiming the new liability rules issued by Secretary Hickel were confiscatory and a breach of contract. Edwin Pauley, the company president, was instrumental in persuading the Congress to enact a tidelands oil bill in the early 1950s, and his company, presumably, was privy to detailed information on the channel bottom. The complaint filed in the Pauley suit says what the citizenry of Santa Barbara had been saying all along:

> The United States also knew or should have known at this time, as plaintiffs did, that the Santa Barbara Channel area was characterized by deeper water, greater tectonic activity, a greater density of subsurface faults and fault zones, and more frequent and more intense earthquake and other seismic activity than most, if not all, other areas in which offshore exploration for and production of oil and gas had theretofore been attempted and that each of these conditions increased the likelihood of well blowouts, pipeline breakage, and other causes of inadvertent spillage or seepage of oil.
> The United States likewise knew or should have known

at this time, as plaintiffs did, that it was generally understood in the petroleum industry that the possibility of well
blow-outs is, roughly, inversely proportional to the available knowledge with respect to subsurface geologic conditions and that the geologic knowledge of the subsurface
outer continental shelf areas under the waters of the
Santa Barbara Channel was even more sketchy and un-
certain than the geologic knowledge available with respect
to on-shore oil-producing areas and most other submarine
off-shore oil-producing areas in the United States.

Thus, when the United States solicited bids and substantial cash bonuses for oil and gas leases under the
waters of the Santa Barbara Channel, it knew or should
have known, as plaintiffs did, that no operator could
guarantee that, even with the greatest degree of care, its
exploration and production in the area would be free of
well blow-outs or of other events which would give rise to
the unintended discharge of oil into the surrounding
waters.

Pauley went on to claim the leases "are in deep water
and in a known area of faulting which is subjected, from
time to time, to earthquakes and tidal waves. Drilling in
the leased areas requires operations which reach to the
presently known limits of the relevant technology."

There is an ironic twist to the OCS leasing policies.
Some of the OCS royalty money goes into the Land and
Water Conservation Fund. This fund, established in
1965, is used by the Interior department to purchase
lands for national parks and other preserves. In 1968,
the conservationists worked hard to push through
amendments to the act doubling the size of the fund to
$200 million. Until then, $100 million came from reve-

nues received from fees paid by tourists to national parks, a motorboat fuel tax and proceeds from sale of surplus lands. In 1968, the Congress agreed to double the fund to $200 million a year, stipulating the additional funds come from revenues received from leases to the outer continental shelf. In this way, conservationists tied themselves to the idea that some land could be saved by plundering other land. Money received from the Santa Barbara leases went to help save the Indiana Dunes, preserve parts of the Atlantic Coast near the Chesapeake Bay, and buy more land for the Point Reyes park north of San Francisco.

The amendments to the Land and Water Conservation Fund illustrate a continuing theme at Interior: conservation proceeds apace with exploitation.

Following the Santa Barbara spill, California Senator Allan Cranston and Representative Charles Teague, a liberal and a conservative, sponsored legislation to halt all drilling in the channel and give the companies back their money. Significantly the bill never received support from the water-pollution crusaders, Muskie and Blatnik, and lies buried in the Interior committees, always so partial to the oil interests. As for the new stiff safety regulations set into operation by Hickel as a result of the Santa Barbara spill, they are routinely ignored by industry and the Geological Survey which is meant to enforce them. On one wild well in the Louisiana Gulf, there were more than 300 safety violations. Most important, the government quietly prepares to issue leases on two large tracts of outer continental shelf off Louisiana and Alaska.

The entire energy industry, dependent as it is on the operations of the big petroleum companies, is grossly expensive, inefficient and wasteful. The oil markets must continually be juggled by the big American-based firms in order that they may shuttle oil from their different fields to different markets. At base, the system depends on making sure there is impetus to drill for oil within the continental United States. A high rate of US production is absolutely necessary for establishing a big tax write-off, which, in turn, is essential in maximizing profits.

6 | Masters of Waste

Whether or not President Nixon's war on pollution was concocted as a deliberate measure to draw attention from the Vietnam war, which it succeeded ever so briefly in doing, it is not likely to make many changes for the better in controlling environmental pollution. In fact, in some respects, Nixon's program is likely to increase the dangers of pollution. His efforts are led in large part by Secretary of Interior Walter Hickel, the former governor of Alaska, with close ties to the oil industry. His "Clean Water Team" consists of Carl Klein, a Chicago savings-bank lawyer and friend of the late Everett Dirksen, who is Assistant Secretary at Interior, and David Dominick, a thirty-three-year old cousin of Peter Dominick, the Colorado senator.

Not long after he took office, Hickel gave Edgar Speer, president of US Steel, a special clean-water award in recognition of US Steel's "initiative in pollution abatement practices at the company's Lake Superior, Lake Erie and Lake Michigan facilities." At the ceremony Hickel declared, "What US Steel has done is both good policy and good industrial practice. It is making a positive contribution to assuring adequate water supplies for future industrial and domestic expansion," and added, "The water re-use philosophy of US Steel must become widespread if the industrial areas of the Great Lakes are to continue to provide jobs and income for millions of people."

Subsequently the government sued US Steel as one of the biggest polluters in the nation. In May 1970, US Steel's South Works on Lake Michigan was found guilty of contempt of court by a Chicago judge because it had been polluting the lake with oil in violation of an injunction issued against such practices two years previous.

Several weeks after the Speer award, the Clean Water Team threw a $100-a-plate party at the Washington Hilton for 600 executives who had dedicated themselves to the sanitation crusade. Speer set the tone: "We oppose treatment for treatment's sake," he declared, pointing out there just were not enough earnings for "ideal" pollution abatement programs. "Unless the money for pollution control is intelligently spent—not by the dictates of emotion—the citizen is paying for something he didn't get," Speer said. "Is an additional 10 percent improvement in fishing worth $100 million?" John Swearingen, chairman of Standard Oil of Indiana, added, "The central question is not whether we should have cleaner water, but how clean, at what cost, and how long to do the job. These considerations are frequently ignored in popular discussions. Public enthusiasm for pollution control is matched by reluctance to pay even a modest share of the cost. This attitude will have to change."

Klein believes that the states, not the government, should handle the pollution problem. He and Dominick announced a re-organization of the pollution-control administration involving the rotating of regional directors. This allowed Klein to shift H. W. Poston, the regional director in Chicago with a reputation for being tough with industry. Poston refused to rotate and quit instead. In his place Klein proposed Val Adamkus, a freelance engineer, who came to the Administration's attention as chairman of a Chicago Lithuanian Citizens for Nixon group. That created a furor in the Chicago papers, and Adamkus became a deputy regional director in Cincinnati instead.

As part of the Nixon drive against pollution, emphasis shifted from enforcement conferences to grand-jury inquiries. In New York and New Jersey, grand juries handed up criminal informations against certain industries charging pollution under the Rivers and Harbors Act of 1899. Convictions on such an information could

at most bring a fine of $2,500 per instance. The shift in
enforcement away from the pollution statutes, so labori-
ously legislated over half a century to the Rivers and
Harbors Act, brought into the open a quarrel between
the Justice Department and the FWQA. In Chicago,
assistant US attorney, Jack B. Schmetterer, said criminal
prosecutions were hampered by refusal of the Interior
department to provide assistance: "The one federal
agency that has sophisticated technical staff and information sufficient to help us move forward against major
water problems in the Northern District of Illinois has
refused to give us that aid. When I say 'regret' I mean
it because any concept of professional Government service means inter-agency cooperation to attain important
public goals, and we usually have that cooperation in
the Federal Government. But the absolute refusal of the
Interior Department to permit the Federal Water Pollution Control Administration [now FWQA] office to
supply us *any* information or technical advice defies
understanding. Our request for specific information on
specified companies has gone unanswered. Our request
for general advice and judgment as to which companies
pose the most critical problems has gone unanswered.
Our final request for just such information as would be
made available to any member of the public upon request has not been complied with. Apart from a few
published reports and transcripts, we have received
nothing. The technical information and advice held by
Interior Department and made available to the states is
not available to the United States Attorney. It has

proved difficult even to get the FWPCA laboratory to analyze samples picked up by Coast Guard, and we have asked the Guard to return to its former procedure of delivering samples to the Corps of Engineers for a time-consuming analysis by a private laboratory.

"Agency cooperation is absolutely necessary to an effective prosecution program by our office. We are chronically understaffed, spread thin with criminal and civil responsibilities in every conceivable field of federal law, and face an incredible trial and appeal case load. Many assistants average up to 70-hour work-weeks already. We therefore lack time and facilities to do basic research in technical subjects and search nation-wide for our own experts, as we now must. Therefore, lack of Interior Department assistance has made much more difficult our effort to analyze pollution by major offenders through the Grand Jury process.

"Facing this lack of cooperation we tried to persuade the agency to change its views. We have tried for several months to work quietly to bring about that change in attitude. Perhaps that attitude is due to the view of an Interior Department official who called me from Washington to ask why I wanted to prosecute 'those nice people.' Perhaps it is due to other reasons. Whatever the reason, I decline to keep silent on the problem any longer, and have spoken out in the hope that public interest will insure good coordination of the federal effort in this vital field. It is an area too important for bureaucratic wrangling over jurisdiction or political fighting over who gets the credit. This is the public interest we are about, and if we are really a law-enforcement minded country, the pollution laws are among those which must be enforced."

After Chevron lost control of ten wells in the Louisana Gulf, Hickel discovered the company had violated department regulations for oil operations in 300 different instances. On March 25, he asked the Attorney General to convene a grand jury in New Orleans to investigate allegations that Chevron Oil Company had violated regulations. (The company was subsequently indicted.) Hickel said he hoped the violators would be "prosecuted to the full amount of the law." But when he was reminded that the law provided for cancellation of a lease where there was a violation of regulations, Hickel said he had not contemplated any penalty beyond a fine ($2,500 a day). However, he did postpone leasing more underwater tracts on the outer continental shelf until he could study the situation.

Shortly thereafter, newspapers ran photographs of Vice President Agnew emerging from a New Orleans hotel surrounded by oil men. Agnew had met with them to discuss the oil spill. On April 4, the Associated Press reported Hickel had advised Louisiana Senator Allen Ellender that he soon hoped to resume oil-lease sales in the Gulf.

The practice of fining large companies for pollution is a dubious proposition. The result may be to legitimize pollution, since in effect it charges them a fee for discharging filth into the water.

The Nixon Administration has repeatedly fought

efforts by the Congress to increase spending for water-pollution programs. In 1969, Klein told the House appropriations committee there was no need to appropriate more money for sewage grants; he maintained the pollution headquarters had money left over from 1968. But when the House insisted and voted $800 million, instead of the $214 million Klein wanted, the Assistant Secretary said they would just have to live with the increase. The Senate voted the full $1 billion called for in the authorizing legislation which was compromised to $800 million on conference. The President signed the bill reluctantly. But the Budget Bureau told the FWQA the $600-million difference between what Nixon wanted and the Congress voted would not be spent. Meanwhile, Nixon was pushing his publicity war on pollution. Eventually, he agreed to spend the money appropriated.

Although Nixon declared war on pollution, he apparently will spend less money than any president since the Federal program was begun a decade ago. The President said he believes it necessary to spend $10 billion over the next five years to clean up water. Of that total, the government would pay $4 billion: local communities would be required to come up with the remaining $6 billion. Federal grants for sewage works are paid out over a nine-year period, so this is not as large an amount of money as it may at first seem. In fact, annual appropriations would come to about $450 million, less than the $800 million appropriated this year. Some of the $445 million will go to pay states for

sewers they have already built, and for which the
Federal government has promised them grants-in-aid.

There is yet another side to the economics of the
President's sewer program. As it turns out, the Nixon
program looks like a special pork barrel for the northeast states.

Since the first sewage-treatment plant was built in the
United States in 1856, the government has invested $9
billion in plants which treat sewage from 92 percent of
the population. Now in four years, Nixon proposes to
commit $10 billion to treat the other 8 percent of the
population (10 million people) and to bring all treatment levels up to the secondary stage. About half of the
money will go to the northeast and that includes Pennsylvania, New York and New Jersey, as well as the New
England states. The highest per capita expenditures
will be in Maine, New Hampshire and Vermont, states
which lag far behind in building sewage systems. They
are largely rural communities so that building sewers
and attaching waste-treatment plants will be expensive.

Probably the least expensive way to effectively sewer
these rural communities would be through septic tanks,
lagoons or some form of Kardos' "living filter" concept
(another version of Chadwick's idea). But the sanitary-
engineering fraternity and public-health doctors deride
septic tanks and lagoons as health menaces. They insist
communities should be sewered and their wastes treated.

Naturally sewage-treatment plants cost much more
than a lagoon or septic tank, and involve considerable
study by consulting engineers and much heavy con-

struction work. According to the Federal Water Quality
Administration, it costs on a national average $179,000
to construct an activated-sludge plant which handles
10 million gallons a day. But in the northeastern states,
that plant costs $441,000, or 246 percent more than any-
where else in the United States. This is partly due to the
enormous profits of the construction industry, which
runs 13.7 percent after taxes. (That is more than double
a 5 percent profit which is usual in manufacturing industry.) On top of that the consulting engineer takes
out his 6 percent. This is an immensely lucrative business, and with more and more money scheduled to be
spent for sewers and sewage-treatment plants, it could
well become one of the biggest pork-barrel items in the
Federal budget, rivaling the already infamous highway-
building program.

Looked at in this way, the Nixon program is pork
directed to the engineering and construction industry in
the northeast, through the governors, three of whom are
prominent Republicans—Shafer in Pennsylvania, Cahill
in New Jersey, and most important Rockefeller in New
York, who has been pumping money into sewers as if
he were running Tammany Hall.

Even if the money were equitably distributed and
spent efficiently, it would be without much meaning,
for the major cause of pollution is industry, not human
sewage treated through municipal waste-treatment
plants which are eligible for Federal grants. For every
one pound of BOD from a person, there are four pounds
from manufacturing industry. There are 30 trillion

pounds of industrial BOD per year compared to 8.5 trillion pounds of human BOD. Industrial growth has produced a terrific demand for water. For example, it is said that 18 barrels of water are required to refine a barrel of oil; 300 gallons of water to make a barrel of beer, 600 to 1,000 tons of water for each ton of coal burned in a steam-power plant, 250 tons of water for each ton of paper, and that a large paper mill will require more water than a city of 50,000 persons.

By advocating a spending plan directed to the northeast, Muskie and Nixon sidestep the principal issue of industrial waste. However, both men do propose an approach.

They would encourage the building of new regional sewers and sewage-treatment plants which could accommodate both industrial and municipal sewage. Where industries already are connected to city sewers, then Nixon proposes to upgrade the processes to the secondary-treatment level. However, half of all industrial wastes are inorganic, mostly primary metals, and the biological sewage treatment afforded by either activated sludge or some form of filtration will not handle this sewage. The industrial sewage is likely to poison the whole system, and ruin whatever effectiveness it had.

The current fad for replacing small waste-treatment plants with big, centralized regional systems is good business for construction industry and the engineers, but it is not necessarily an efficient way to deal with sewage. Older, smaller plants may work just as well as

one great big plant, especially since the technology has
not changed since the turn of the century.

If an industry joins in a regional sewer system, then
the whole system can qualify for up to 55 percent of the
cost in Federal grants. In addition, under the Nixon-
Muskie plan, if the community is unable to sell bonds,
on the public markets, then they will be purchased by
the government at below market interest rates. That
means the general citizenry will pay for treating industrial sewage. Muskie, however, wants to further reduce
the pressure on industry by increasing the Federal share
of the cost to 65 percent. States would contribute 25
percent, and the local communities would contribute
only 9 percent. That means the burden on industry
would be substantially reduced.

Nixon has made other benefits available to industry.
Under the tax-reform act of 1969, there is a provision for
five-year rapid write-off on pollution equipment. The
Federal government will pay up to 70 percent of the
costs of industry's research programs aimed at developing pollution-abatement schemes.

Part of the problem in effectively dealing with industrial pollution is that neither state nor Federal governments has substantial information as to the amounts and
types of industrial pollutants. In 1964, Congressman
Bob Jones of Alabama requested the Secretary of Health,
Education and Welfare to compile an industrial in-
ventory, but the idea was quashed at a meeting of Federal officials at the Budget Bureau. Jones tried on two

other occasions to get the questionnaire approved by
Budget, but met with no more success. In May 1970,
proposals for a questionnaire to industry were again
stuck at Budget. Meanwhile, Nixon appointed a National Industrial Pollution Control Council, made up of
fifty-three businessmen, which will advise him on all
matters pertaining to industrial-pollution reform. The
chairman of the group is Bert S. Cross, head of Minnesota Mining and Manufacturing, a company which
has been under orders since 1967 in Wisconsin to stop
discharging sulphur wastes into the Mississippi.

While Nixon was publicizing his pollution campaign,
former Secretary of Health, Education and Welfare
Finch testified against Senator Muskie's bill, which
would create a pilot program to test means of recovering
and re-cycling solid wastes such as plastic, paper, tin cans
and bottles. The bill would authorize a research program, and make available some money for experimental
projects. Finch said it was too costly.

The Nixon Administration lobbied against "absolute
liability" provisions in amendments to pollution legislation. Under the Senate version of a bill sponsored by
Senator Muskie (see Chapter 5), shipping companies
would be responsible for cleaning up all oil spills up to
$16 million unless they could demonstrate the spill was
caused by an act of war, God, or by a third party. Under
the House version, sponsored by Congressman Blatnik,
the government would have to prove gross negligence to
prove liability. The House version reflected the interests
of Congressman Cramer, a Republican of Florida.

Nixon backed his version. Gross negligence is extremely difficult to prove and the phrase makes the law meaningless. After three oil spills in Cramer's district, he gave in, and the Senate version passed conference. Nixon signed the legislation.

Although Congress passed water-pollution legislation in 1965 which directed the government to establish national water-quality standards by 1967, those standards still have not yet been set. This is an embarrassing situation and in October 1969, David Dominick promised to resolve the outstanding exceptions by states to the standards by January 1, 1970. However, by mid-January, the government had approved complete sets of standards for only four states—Arizona, Minnesota, Nebraska and Utah—and two territories—the Virgin Islands and Guam. By mid-May, the list of approved states had increased to twenty. Even where the standards are in effect, implementation dates range from 1972 to 1980.

Meanwhile, the crusaders for water-pollution control in the Congress are finding themselves in equally embarrassing situations. Muskie advocates water-pollution reform, but he also is lobbying for an oil-refinery complex at Machias, Maine. Such a complex would result in increased super-tanker traffic off Maine, and because of the shape of the coast and the roughness of the Atlantic, the project poses more than the usual pollution dangers. In addition, the refinery itself, with chemical industry attached, could create pollution.

After asking the General Accounting Office to make a study of the effects of pollution legislation, Muskie

swept the report under the rug. That report, as indicated earlier in this book, seriously questioned the
worth of the entire pollution program. In particular, it
implicated Muskie with an unfortunate pollution situation in Maine—the Vahlsing sugar-beet factory. More
important, he has been so closely involved in writing
and interpreting pollution laws, that it must be difficult
for him to lead an investigation which might show the
laws do not work. Given his political ambitions, that
would be a suicidal course.

For his part, John Blatnik, the Democratic congressman from Duluth, Minnesota, who chairs the House
Rivers and Harbors subcommittee, always has strongly
supported development of taconite mining as a way to
create employment in the depressed Mesabi iron-ore
Range. Now taconite mining, which involves using immense amounts of water, has become a major source of
pollution on Lake Superior. The Federal Water Quality
Administration has begun enforcement proceedings
which could result in curtailing taconite mining. Blatnik is left very uncomfortably in the middle.

Nixon's proposals for strengthening the air-pollution
program would almost surely make the national campaign more ineffective than it already is. Until recently,
the government's air-pollution program, based in the
National Air Pollution Control Administration within
HEW, has not amounted to much more than a public-
relations effort directed at awakening citizen interest. In

1967, however, at the urging of Senator Muskie, Congress passed amendments to the pollution act which established so-called "ambient" air standards on a regional basis. Under this scheme, the government developed criteria which were meant to assist the states in establishing standards. The idea was that air in a given region should not contain more than certain specified levels of pollutants. The states were somehow supposed to set standards to achieve this. This idea posed an enormously complicated, probably impossible task. A more direct way to deal with air pollution would be to set emission standards for stationary sources, such as utility smoke stacks, and for pollution contents in fuels. Industry opposes such a course, however, and the Air Quality Act of 1967 was widely viewed as a victory for industry.

Now Nixon proposes to make matters more chaotic by directing the Secretary of HEW to establish national ambient air-quality standards, which would be enforced by the states. This idea incorporates the poorer aspects of the 1967 Muskie legislation. The whole concept of ambient air-standard setting remains virtually incomprehensible. And by leaving pollution enforcement in the hands of the states, Nixon would ensure that industry had a strong say in what happens.

Moreover, Nixon wants to cut back funds for air pollution. Last year he sought to reduce the key research funds from $38 to $27 million. That money goes for research on developing emissions controls for vehicles, and for work on building alternate engine systems. The President also attempted to frustrate the existing feeble pro-

gram for developing alternate engine systems. Senator
Magnuson had sponsored legislation which would create
a $50 million fund for procurement of vehicles powered
by new kinds of non-polluting engines. Every year the
government buys 60,000 motor vehicles, and the idea
is to create a market for pollutionless cars, trucks and
buses within the government fleet. Under Magnuson's
plan, the government would pay the manufacturer of a
pollution-free engine 25 percent more than he would
normally get. The Nixon Administration attempted to
ruin the idea by insisting that purchases of pollutionless
vehicles be restricted to automobiles. Since automobiles
account for but 10 percent of the government fleet, that
would render the Magnuson bill practically worthless.

While the Nixon policies on air and water pollution
work to legitimize and spread pollution, the Administration pursues policies aimed at exploiting natural resources in other areas as well.

   Under the mining law of 1872, anyone may enter Federal lands, except national parks and other areas specifically closed by law, to prospect for minerals. (The mining law pertains basically to metals: iron, copper, lead,
uranium, etc. Coal, gas, oil phosphates, sulphur are covered by a separate leasing law.) An individual may
stake a claim of twenty acres, or up to 120 acres if joined
with other people. The claim is filed with the county.
Although the land is Federal, the government has no
knowledge of a claim being filed. So long as the individ-

ual or company searches for minerals, he can file for a
patent to the claim. On paying a fee of $2.50 per acre
for a placer claim and $5 per acre for a lode claim, the
government must give the prospector full title to the
land, including title to the surface resources; which often include valuable timber. The fees were established
in 1872: there are no regulations as to how the miner
goes about his business. In exploring his claim, a prospector may cut a road through timber, bulldoze, dredge
or strip at will. In fact, in some of the Western states,
prospectors are encouraged to gouge the land. State laws
require they dig a pit and throw up a works of sorts on
each claim as proof they really are working it. In Wyoming, for example, mining companies bulldoze or
plough big strips of land in search of jade, and then
move on to new land without any attempt to restore
the acreage they have just ravaged. In Arizona, where
half the United States supply of copper is produced, big
open pit mines outside of Tucson use so much water
the groundtable has dropped, threatening a drought in
the city. Great cliffs of mine wastes called tailings are
built up around the mines. The wind blows the wastes
into neighboring towns, causing severe air-pollution
problems. The uranium companies have left big piles of
radioactive tailings along the Colorado River, where
they endanger the water supply. In Southern California,
the 1872 law has become a subterfuge for acquiring
land. According to the Interior department, some 2,000
claims have been filed there in recent years by individuals seeking public land for a residence or camping

site. If the government wants to dislodge one of these "miners," it must proceed case by case, taking each one separately.

In the last days of his tenure at Interior, Udall urged the mining act be changed around, in effect switching minerals over to the lease basis. But Hickel continues to support the old act. In a letter to the Public Land Law Review Commission, which is trying to decide what the policy toward public lands should be, Hickel said, "It [the mining law] is not the villain that it is so frequently portrayed. Perhaps the most important function this law performs is the stimulation of individual incentive to seek out and develop valuable minerals which are essential to the continued growth and prosperity of this nation."

Hickel goes on to explain what he means by this: "Most criticism of the mining law has centered on its failure to provide sufficient protection for ecological factors and on the fact that the federal government receives no direct compensation for minerals discovered and taken. It is my view that answers to the valid objections can be found, that all the varying interests can be reconciled and that we can devise a workable revision of the mining law of 1872 which will enable us to meet our present and future needs. This can be accomplished without sacrificing the best qualities of the old law while still insuring appropriate consideration for necessary conservation and multiple use management."

Nixon's message on the environment was especially noteworthy in its absence of any mention of mining or the mining laws. The 1872 law remains in force, providing a free hand-out of land to the big mining interests.

Each year the Secretary of Interior rents 100 million acres of public grasslands to ranchers for grazing. The cattlemen now pay 44 cents per animal unit month. The price is cheap since privately owned grass goes for as much as $3.50 to $4 per animal unit month. While this is meant to subsidize the little rancher, it doesn't work that way. Much of the Federal rangeland is taken over by big cattlemen, and they lobby vigorously to keep the price of range grass low.

The government sets aside some of its revenues from renting rangeland for re-seeding it. The range has been going downhill since the First World War. According to the Interior department's own estimates, some 30 percent or 50 million acres of the range is in bad shape. By refusing to raise the range rent, Hickel is ensuring that that there won't be enough money in the range-improvement fund to re-seed worn-down grass. According to estimates by Montana Senator Lee Metcalf, Hickel will be foregoing treatment on some 150,000 acres annually, and thereby laying open the ranges to increased erosion.

Nixon's policy of despoilation of the natural resources is not limited to the Interior department. In early 1970, he sent the secretaries of Agriculture and Housing and Urban Development to the Congress to plead for enact-

ment of the so-called National Timber Supply bill. That
measure would allow lumbermen to increase their already large cuts of publicly owned forest lands. Nixon's
men argue that timber is in short supply for housing,
and the lumber companies should be permitted to cut
down more trees. Actually, Nixon's own economic policies have dragged the home-building business to a stand-
still, and at any rate there are plenty of other building
materials.

Finally, one of the most serious obstacles to any sort
of sensible, national conservation plan, is the doddering
fleet of United States merchant ships. Under the cabo-
tage law, the Jones Act, all water-borne commerce among
the states of the Union must be carried in US bottoms.
Because operating costs on US ships are so much higher
than on foreign-flag vessels, the effect of the Jones Act
is to encourage shipment of US natural resources to
markets abroad on less expensive foreign-flag ships.
Thus, Alaskan natural gas, badly needed in West and
East Coast cities where pollution is severe, is sent instead to Japan. The same is true for Alaskan timber.
East Coast cities are encouraged to import gas from Al-
geria. The effect of the Jones Act, which is staunchly
supported by the shipping industry, is to deny commerce
among the states, not to foster it.

Matters are now likely to get much worse because
Nixon has embraced the US shipbuilders, and promised
the industry $3.8 billion in subsidies over the next ten
years. That is the largest subsidy ever granted. The US
shipping lines now will be encouraged to build super

tankers to haul the oil from Alaska and the outer continental shelf. It means prices will go higher and, in all likelihood, the big petroleum companies will be offered an excuse to keep oil out of the US, and instead send it abroad.

THE NEO-MALTHUSIAN CRUSADE

Efforts to cope with pollution have been sidetracked by the reappearance of the Neo-Malthusians. They insist the basic cause of pollution is not industry, but rather the increasing population of the United States, along with a rising flood of humanity abroad. (Of course many people besides neo-Malthusians want to limit population and for a variety of reasons. But the neo-Malthusian idea historically has dominated the population control movement.)

Well-to-do white people have had a passion for population control since the eugenics movement of the early 1900s. The eugenicists were influential in establishing the immigrant-quota system, designed to stop the influx of southern and eastern Europeans. Henry Fairfield Osborn, a naturalist and one of the founders of the American Museum of Natural History, brought the second International Congress of Eugenics to New York in 1921. He argued, "The moral tendency to the hereditary interpretation of history . . . is in strong accord with the true spirit of the modern eugenics movement in relation to patriotism, namely, the conservation and multiplication for our country of the best spiritual, moral, intellectual and physical forces of heredity. . . . These divine forces are more or less sporadically distributed in all races . . . but they are certainly more widely and uniformly distributed in some races than in others."

A nephew, Frederick Osborn, in 1940 believed "from his studies of population decline that it is highly necessary for 'privileged' individuals to produce more children." With the advent of European fascism, the Osborn family backed off the eugenics line. However, they remain very much concerned with population control in its more progressive, humanitarian version.

William Vogt, the ornithologist who died in 1968, was perhaps the best known of the Neo-Malthusians and a tireless advocate of the cause. He worked with population and conservation groups to unite them in a program for population control. Vogt wrote *Road to Survival*, in 1948, and influenced the views of Fairfield Osborn, whose book, *Our Plundered Planet* was an immensely popular depiction of the Neo-Malthusian idea.

Vogt began as curator of the Jones Beach State Bird Sanctuary in the early 1930s, then worked for Audubon clubs and was a consultant to both the Peruvian and Chilean governments on ecological matters. After the Second World War he was made chief conservationist for the Pan American Union. He was president of Planned Parenthood and director of Laurence Rockefeller's Conservation Foundation. Because of his work in Latin America, Vogt became convinced population control was the chief ecological interest. In *Road to Survival* he argued that population was fast increasing while natural resources were being ruined. As a result people were starving: "The angry muttering of mobs, like the champing of jungle peccaries, is a swelling echo of their passing." Overpopulation leads to war because nations in search of more land to feed their populations must expand. The invention of the diaphragm was a major contribution to world peace, but other devices were needed. Sterilization bonuses would be a step in the right direction. "Since such a bonus would appeal primarily to the world's shiftless, it would probably have a favorable selective influence. From the point of view of society, it would certainly be preferable to pay permanently indigent individuals many of whom would be physically and psychologically marginal, $50 or $100 rather than support their hordes of offspring that both by genetic and social inheritance would tend to perpetuate their fecklessness." As for proffering aid to Europe, "We are in a position to bargain. Any aid we give should be made contingent on national programs leading toward population stabilization through voluntary action of the people."

During the 1950s the Neo-Malthusian line achieved increased importance when the Rockefellers put money into the population-control movement, by financing the Population Council, Planned Parenthood and the Population Reference Bureau. In 1957, an ad hoc committee of population experts from the Council, the Rockefeller Fund, Conservation Foundation and Planned Parenthood published a scheme for controlling populations called "Population: An International Dilemma." The report said population was the key to stability in both rich and poor nations. The idea was to persuade educated people of the population dangers. Birth control itself would grow out of the dictates of family planning. The committee believed population was a problem in the United States. "Excessive fertility by families with meager resources must be recognized as one of the potent forces in the perpetuation of slums, ill-health, inadequate education, and even delinquency." The committee, however, believed that there was an overall "balance of population and resources" in the United States and sought only to use tax, welfare and education policies "to equalize births among the socially handicapped." In the 1950s population groups urged the government to extend development aid to local maternal and child-welfare programs, research and to groups studying the problem of population. Eisenhower himself would not touch birth control, but Kennedy agreed to a US role in research and offered assistance to the United Nations.

While the US does not spend much money to spread birth-control information and materials abroad, population control is nonetheless a major tenet of foreign policy. It is essential to the accepted US theory of industrial development in underdeveloped countries. Stephen Enke set the line in this area with a paper published in the *Economic Journal* in 1966. Enke wrote: "With crude birth-rates continuing at 40 plus per thousand a year and death-rates continuing to fall, many nations' population at present rates of natural increase will double every 25 to 35 years. Perhaps the employed labor force can double as fast. But natural resources cannot increase by definition. And many poor countries cannot save and invest enough yearly to double their stock of capital in, say, 30 years. Therefore, unless innovations increase final output to factor input rates rather more rapidly than now seems the case, aggregate output per capita may barely increase. Most of these countries cannot both have natural increases in population of from two to three percent annually and increases in per capita income of three percent a year or better. . . ."

Enke went on to argue, "If economic resources of given value were devoted to retarding population growth rather than accelerating producing growth, the former resources could be 100 or so times more effective in raising per capita income in many less developed countries. An adequate birth control program in these countries might cost as little as 10 cents per capita yearly, equivalent to about one percent of the cost of current development programs. The possible use of bonuses to encourage family planning, whether paid in cash or kind, is obvious in countries where the 'worth' of permanently preventing a birth is roughly twice the income per head."

The Enke theory is litany around AID and in the halls of the World Bank, where Robert McNamara, the president, has adopted the population-control cause.

In 1963, the population-control people created an organization called the Population Crisis Committee,

and made General William Draper chairman. It functions as a lobby on two levels attempting to push more
US funds into birth-control programs administered by
AID: acts as a quasi-governmental organization through
which US funds can be routed to birth-control programs
in underdeveloped countries. Since the US finds it embarrassing to funnel money for birth control to backward Catholic countries, it runs the money through
Draper's group and from there it goes out to Planned
Parenthood groups abroad.

Recently the committee expanded its work, and increased the staff. James Riddleberger, former ambassa-
dor to Austria, became executive director. Riddleberger
was the first administrator of the AID program when
it was launched in the late 1950s by Eisenhower, and he
was a proponent of family planning even then. Ernest
Gruening, the former Democratic senator from Alaska,
is a consultant. Gruening ran non-stop population-control hearings while he was senator. Riddleberger and
Gruening, with their contacts in the State Department
and on Capital Hill, are meant to break open the government coffers and finance birth-control programs. The
most notable victory for the Draper group so far has
been Nixon's warm endorsement of population control,
complete with Neo-Malthusian rhetoric.

Essentially, the people around the Rockefeller population movement are older, relatively conservative types,
nostalgic for an earlier time. They find the management
of the imperial domain has come to a sordid pass.

Within the last five years, the population groups have

split apart. The conservatives, who cluster around Planned Parenthood, stop at giving advice to women on planning their families. Other groups, such as the Association for Voluntary Sterilization, have moved to propagandizing for the two-child family, freely giving out information, devices and engaging in ambitious advertising schemes for population control. Paul Ehrlich, the Stanford professor and author of *The Population Bomb,* has galvanized the change. He has worked hard to bring radicals in the population groups together with conservationists. Ehrlich sees the population explosion breaking down the entire ecological system. He wants to find prompt political solutions, and he urges institution of a tax on families with more than two children. He believes that extension of foreign aid should be tied to an acceptance of birth control by the recipients.

Ehrlich has discussed the possibility of placing sterilants in the water supply to stop off the flow of children, but he concedes they pose a technical problem, which as yet is unsolved. However, he believes it would be useful to add special luxury taxes on layettes, cribs, diapers, diaper services, expensive toys, etc., always with the proviso that such services be supplied free to the poor. In addition, the government might award a "first marriage grant" to people who got married after they were twenty-five. "Responsibility prizes" might be handed out to couples who went five years without having a child, or to a man who turned himself in for sterilization. And there could be special lotteries for prizes and money; only those without children could qualify to buy tickets.

Compared to his colleagues, Ehrlich comes off seeming rather restrained. Dr. Garrett Hardin, a biologist at
the University of California at Santa Barbara, says,
"Freedom to breed will bring ruin to all. . . . The only
way we can preserve and nurture other and more pre-
cious freedoms is by relinquishing the freedom to breed,
and that very soon." According to Donald Aiken, a
physicist, "The government has to step in and tamper
with religious and personal convictions—maybe even
impose penalties for every child a family has beyond
two."

The publication of Ehrlich's book brought the population movement together with conservation groups.
The Sierra Club, in conjunction with Ballantine Books,
actually published *The Population Bomb*. While the
Sierra Club does not endorse Ehrlich's proposals, it is
taking an increasing interest in population matters. So
does the National Audubon Society. Zero Population
Growth, Inc., better known as Z Pop, was begun in 1969
with Ehrlich as the chairman. Z Pop pushes for no net
population growth in the United States by 1980 and the
rest of the world by 1990. It wants birth-control information and supplies made available to everyone. That includes tax-supported programs for the poor. Freely available birth-control methods should include all types of
contraception, voluntary sterilization and voluntary
abortion. Z Pop talks about building a congressional
coalition, and to that end, providing members of Congress with interns to work up studies and various legis-

lative schemes. The Z Pop people want a two-child family, and they talk about pushing for legislation which could achieve this end, involving such things as higher taxes for large families.

The Neo-Malthusians insist in arguing that population programs mean economic betterment for the family. In many poor countries, where wealth is concentrated in the hands of a very few or held externally through multinational corporations, that would necessitate redistribution of wealth. But there is little evidence to indicate that population-control programs result in income redistribution. It can work the opposite way around. Population control can be a means for rulers to control the populace. And when the issue is considered in terms of modern technology, population control can become a way to narrow and increase the wealth of a few individuals and corporations. The United States, with but one fifth of the world population, uses 50 percent of the world's resources. Decline in the population of the rest of the world is simply a means of advancing our own interests.

The "Population Explosion" within the United States is not as depressing as the Neo-Malthusians claim. The US has a current population of 205 million. The people are spread over 3,615,123 square miles of land, for a density of 55 persons per square mile. Ben Wattenberg, in the *New Republic,* writes: "As measured by density, this makes the United States one of the most sparsely populated nations in the world. As measured by density, Holland is about 18 times as 'crowded' (at 975 persons per square mile), England is 10 times as dense (588 persons per square mile), scenic Switzerland seven times as dense (382), tropical Nigeria, three times as dense (174), and even neighboring Mexico beats us with 60 persons per square mile. The US, by international standards, is not a very 'crowded' country."

The US has experienced a major re-adjustment of population. In the last eight years one out of three counties in America actually lost population. Four states lost population; two others had a net increase of less than one percent in eight years. Three out of every five counties had a net out-migration.

The fertility rate, an index of the "population explosion," is declining. The fertility rate is the number of babies born per thousand women aged fourteen to forty-four. In 1940, the fertility rate was 80. In 1950, as a result of the baby boom, it was 106, and by 1957 it reached 123. Since 1957, the rate has declined—to 119 in 1960, to 98 in 1965, to 85.7 in 1968. In 1969, the fertility rate was down slightly to an estimated 85.5.

In numbers the population will continue to grow, perhaps to 280 or 290 million in the year two thousand.

However, this population growth does not seem to be related directly to environmental pollution. Ehrlich believes that population is the most pressing problem posed by environmental pollution. But it is questionable whether reduction of population in the United States, for instance, would do much to alleviate environmental pollution. As Barry Commoner, the ecologist, points out, much of the stress that caused pollution began at the end of the Second World War. During that period there was tremendous growth in per capita production of pollutants. Between 1946 and 1966, total utilization of fertilizer increased by 700 percent, electric power by 400 percent, and pesticides by more than 500 percent. Yet in that period US population increased by only 43 percent. This suggests that technology and industry, not the increased numbers of people, increased pollution. To look at it another way, the most severe population explosion exists in India. But India contributes little or nothing to environmental pollution.

A more fundamental difficulty with Ehrlich, however, as with the other Neo-Malthusian ecologists, is that they do not place their program in any sort of cultural context. Ehrlich's proposals would lead to a greater use of technology in the form of the pill, coil, etc., administered to women, without any great concern for the woman's health. Just as pesticides can break down the environment by killing fish and aquatic plants, so can birth-control instruments break down the environment by causing harmful physical and psychological effects in women.

With the emergence of the Women's Liberation movement, the population game as it has been played for the past half century may be over. While the wom-

en's groups differ, there is a common theme to them all.
They are rebelling at the roles male society directs them
to act out. The single most oppressive feature of male
society is the sex game. A true sexual revolution, as
women's liberation groups see it, deals with the social
and political reasons why women have babies, and not
the physiological. Little girls are taught they will grow
up as mothers: not astronauts, firemen or presidents.
Their function in society is the production of children.
Naturally they are reluctant to limit the size of their
families and thereby deprive themselves of the one job
they feel they can accomplish well. This role is rein-
forced by the inability of women to find fulfilling jobs
in a society which offers them menial work. *Off Our
Backs,* the Women's Liberation newspaper in Washington, D.C., says, "Studies have been done which show that
women who work have fewer children. A Worcester,
Massachusetts, study, published in a recent issue of
Science, established a relationship between family size
and sex-role stereotypes. Women who perceived themselves as more 'stereotypically female' (defined by the
participants in the study as gentle and sensitive as opposed to rational and mature—the male stereotype) had
more children.

"We are not proposing that women should not have
children, but that they should have a choice of whether
or not to have a child and how many to have.

"Basically, what we are proposing is this: When
women are allowed greater diversity of roles and when

they are allowed to pursue their own lives irrespective of their choices about marriage and children, then they will not feel it is essential to have many children and a humane system of population control can develop."

It is possible that Neo-Malthusians can bring about through AID and other international schemes some measure of birth control in underdeveloped countries abroad. However, in the United States there is a spreading revolt by women against technology which orders their lives, and directly endangers their health. That revolt could throw off base the consumer industries where growth is based on planned steadily increasing consumption by docile housewives.

A more cynical explanation for the population-control campaign is provided in the lavish newspaper advertising spreads sponsored by the Campaign to Check the Population Explosion. The Campaign is backed by most of the population leaders. One ad says, "How many people do *you* want in your country?" The cities are "packed with youngsters—thousands of them idle, victims of discontent and drug addiction." It says, "You go out after dark at your peril." The solution: "Birth control is an Answer." An answer to what? Racism? Hardly. Another advertisement printed in March 1969 says in large letters, "This is the Crime Explosion." Beneath the slogan is a graph showing crime rates. Then, "And the population explosion is an Underlying Factor." There is a reference to the murders of Martin Luther King and Robert Kennedy. The "quality of life" is declining, and the ad goes on, "Youngsters account for almost half the crimes. And in a few short years millions more of them will pour into the streets at the present rate of procreation." The answer: "Birth Control tackles the problem at its source. A fraction of the vast amounts we spend for health and welfare if devoted to birth control would check the population explosion." Finally, the ad which appeared in June 1969: "Warning: The water you're drinking may be polluted." The basic cause is the rising population in the United States along with the "rising flood of humanity abroad."

The Neo-Malthusian doctrine, rising among both the technocrats and the ecologists, functions as a manipulative scheme aimed at controlling the poor in the interests of the wealthy. The ads are aimed at white, upper-class people who, it is assumed, realize they must control the poor. But it is not the poor who exploited the resources of this continent and turned the waterways into open sewers.

IN "THE PUBLIC INTEREST"
Stripped of the current modish hysteria, the politics of ecology seem dull and complicated, involving groups of quarreling lawyers, doctors, sewermen and industrialists, in the end, exposing the political underbelly of post-industrial America. Behind the banner of Earth Day, competing groups vie for a piece of the environmental movement. I have already described Neo-Malthusians. But there are other layers to the movement.

The ideology for ecology comes from liberal-minded scientists who were first concerned about the dangers of atmospheric testing during the 1950s. Barry Commoner, the well-known ecologist who is a professor at Washington University, was drawn into a study of the environment in the process of trying to chart the effects of radiation. More recently, these scientists have campaigned vigorously against the involvement of the universities in helping the government develop chemical and biological warfare (CBW). They argued against using CBW in Vietnam, and now at home. Until the past few years, Rachel Carson was pretty much alone in criticizing the use of chemical pesticides. A few other scientists supported her position, but the chemical companies led a vicious attack on her, which was backed by scientists from agricultural colleges of the University of California. Within the past year or so, it occurred to the scientific community that pesticides were another form of CBW, and now they have picked up her crusade. The liberal scientific community has contributed information and theory for Ralph Nader's auto-safety campaign, which always has been viewed as a sort of environmental issue.

The ecologists argue in radical, indeed revolutionary, terms for re-organization of society, development of a new political economy which would eliminate ruinous competition, limited production, restructuring of society along more geographical lines, and in general a social order which could enable men to better accommodate themselves to the planet.

Yes, despite the revolutionary rhetoric, the scientists relate more easily to the politics of liberal reform. Beneath the revolutionary rhetoric are arguments for policies which would lead to a more efficiently managed central state, a benign form of capitalism, and they embrace technology as the great problem solver. For their part, the Neo-Malthusians are ambivalent about democratic traditions.

If the scientists provide the ideology for the ecology movement, the public-interest lawyers, financed by foundations and inspired by Ralph Nader, are the movement's political wing. Heretofore, the public-interest lawyers have been identified with consumer protection, but that often involved environmental matters, and now they are moving directly into that issue. The lawyers are the most serious and important part of the movement; they view ecology as an issue which leads to an attack on the power of the corporation. Much of the lawyers' work concerns public health and environmental matters, usually from the point of view of correcting something which is unsafe or inefficient. Their criticism of environmental policies, such as food additives, pesticides, air and water pollution, often begins as an attack on corporate socialism. They want to reform regulatory agencies and believe in the anti-trust ideology. They yearn for the days of the free market.

The most prominent advocates of consumer protection are the trial lawyers who are loosely organized through the American Trial Lawyers Association.

In recent years the trial lawyers have mounted a devastating attack on the American corporation. Their suits begin as simple damage claims, then broaden into attacks on the conduct of the corporation, tracing in the most fundamental, radical manner the injury of individuals back through the corporation hierarchy. They show in detail how corporate officials engage in calculated plunder and injury to the populace. Discovery procedures of the court enable lawyers to obtain and introduce into court records precise details of corporation activity. Juries are convinced and damages running into the millions have been set against companies. Congressional committees assemble and reprint documents from these cases.

The trial lawyers first pursued auto safety. They set up a system for exchanging information on accidents, and concentrated on the Corvair. In the process, they made contacts with scientists and engineers who worked for the auto industry, but had sickened of the job, and who turned over documents detailing the car's faults. Trial lawyers also brought major attacks against drug manufacturers. Where the Congress would not act because of drug-company lobbyists, lawyers attacked through the courts and won a series of impressive victories in cases ranging from Thalidomide to Mer—29.

In the automobile cases, trial lawyers argued that corporations were responsible to the public for every step in the car's manufacture, from the act of designing it on through to the finished product. They insisted that the manufacture of an automobile must be viewed in relation to its environment. The 50,000 people who died on the highway in auto accidents were a "public health" problem. The issue should be viewed in terms of the "pathology" of auto accidents. It was necessary to develop concepts of preventive public health in regard to the automobile, just as it was necessary to do so with measles or typhoid.

Finally, the trial lawyers argued that by producing such autos, the companies were not taking into account the best interests of their stockholders. They were inefficient, backward enterprises. Nader himself argued that by breaking up the industry and creating the conditions for competition, both efficiency and responsibility could be restored.

The effects of this attack by lawyers is profound: the court becomes the legislature. Courtroom victories enhance the power and prestige of the trial lawyers within the legal fraternity as well as among the general public. As in so many other fields, litigation has become a principal avenue for correcting corporate abuse. Even members of Congress, despairing of writing laws, sue in court to accomplish their reforms.

Since the court has become the arena where the post-industrial economy is adjusted, every citizen needs a lawyer—not a legislator—to look after his interests. Perhaps this explains the keen interest by Nader and many other young lawyers in restructuring legal education, improving law schools, developing centers for the practice of law in the "public interest," and so on. In reality they are working toward a new definition of a governmental system in which lawyers are a commanding elite.

At the heart of the ecology movement is the pollution-control industry. Its growth is tied to a series of policies initiated during Robert McNamara's tenure as Secretary of Defense. While McNamara was at Defense, his staff became deeply involved in domestic policy as well as military and strategic decisions. McNamara's theory was that large corporations are best organized to solve the "problems" of housing, crime, unemployment, education and the environment. His assistants argued that if corporations diversified into domestic operations, they would come to rely less on war than peace for profits. The tool for cutting the corporations into social programing would be "systems analysis." The Pentagon wrote key sections of the Poverty Program, which made it feasible for corporations (Litton, IBM, Xerox, etc.) to run Job Corps camps. Through training grants, Defense encouraged the corporations to get into the business of education—to develop teaching products, sell services and eventually to run schools and school systems. The aerospace companies became directly involved in studying pollution problems. As a result they now receive a quarter of the Federal government's research funds for water-pollution control.

While there are different aspects to the pollution-control business, the most important side is devoted to developing additives for controlling pollution once it is created. Companies such as Monsanto, the giant chemical firm which is a major polluter, are excited about the possibilities in pollution because they manufacture pollution-control systems. Future economic growth is absolutely dependent on a continued growth in pollution, along with at least the appearance of rapid technological improvement which will obsolete equipment and create a steady market turn-over for new equipment. Auto companies, which design and manufacture the emission-control systems installed on their vehicles, are in an enviable position of controlling the level of pollution. In addition, they have a device for increasing auto prices. (This can involve sizeable amounts of money. Example: Car prices in 1970 models were increased on average by $107. The auto companies insisted a main reason for the increase was cost of installing safety devices. However, a study by the Labor Department showed the safety devices cost $7.50.) Koppers Company, which makes sewage systems, and systems for removing gases from smoke stacks, also is a major builder of steel-making equipment. Since the steel business expects to expand by 50 percent during the 1970s, Koppers will be in the splendid position of building the new facilities which will increase pollution. Then Koppers can equip the steel works with pollution-control devices. According to the May 1970 issue of *Ramparts,* a count of polluters who have turned to the pollution-control business includes Dow Chemical, W. R. Grace, DuPont, Merck, Nalco, Union Carbide, General Electric, Westinghouse, Combustion Engineering, Honeywell, Beckman Instruments, Alcoa, Universal Oil Products and North American Rockwell.

Another side of the industry is concerned with preventive measures to avoid pollution. A nest of companies works to calm the fears of a community with plans for preserving the environment while at the same time helping them to achieve controlled growth. These organizations function as political brokerages, mediating the interests of competing industries, towns, civic groups, politicians, etc., and representing these communities with state and Federal government. The companies lobby for special legislation, arrange for planning grants, and in general operate as middlemen. Eastman Dillon, the investment banker, provides this sort of function, when it comes to financial matters. Ron Linton, the chief clerk at the Senate Public Works Committee and the person who wrote pollution reforms and managed the air and water-pollution programs for Muskie during most of the 1960s, manages such an environmental consulting business in Washington. Associated with Linton in the business are Hugh Mields, the former lobbyist for the National League of Cities, and Dean Coston, a former Deputy Assistant Secretary at HEW. Both men helped Linton draw up and lobby through pollution reforms. (See Chapter 3.) Another associate is Vernon

MacKenzie, who headed the air-pollution office at HEW while Linton was handling air-pollution legislation for the Senate.

Stewart Udall, former Secretary of Interior under Kennedy and Johnson, is chairman of Overview, an environmental consulting company which helps airports, cities and electric utilities plan for the future. Udall organized the company with Lawrence Halperin, the San Francisco planner. The creation of Overview caused some controversy. While he was still Secretary of Interior, Udall awarded Halperin a big planning grant for preserving the environment of the Virgin Islands. A few months later Udall quit Interior and joined Halperin in the new business. However, Udall maintains Overview never had any involvement in the Virgin Islands deal. Udall says he tried to interest Humble and Atlantic Richfield in hiring Overview to help them plan a sensible program for developing Alaskan oil resources, but the oil companies turned down the offer. They feared Udall's presence would irritate Secretary of Interior Hickel.

Dr. Gordon McCallum, former chief of the water-pollution section when it was part of the Public Health Service, now runs a consulting business. So does S. Smith Griswold, who was air-pollution officer in Los Angeles before coming to Washington to handle enforcement for the National Air Pollution Control Administration.

Perhaps the most important aspect of the new industry are the publicists who are whipping up the market. Here, Arthur Godfrey is the undisputed leader. He arduously campaigns against pollution when he isn't doing TV commercials for Axion, Colgate's polluting detergent. When Godfrey found out he was peddling a pollutant, he handled the situation in a straightforward way. He denounced Axion as a pollutant, and then recommended housewives buy it anyway. Godfrey said Colgate was doing its best to improve the detergent. He claimed it was just not possible to go back to washing clothes with plain soap because there was not enough tallow available, and no matter, washing machines would not work without modern detergents. Environmental Action, that grass-roots movement of student activists who planned Earth Day, provides a sort of Up with People cheer squad for the industry movement. That group was stitched together by Senator Nelson. The top staff was imported from the Kennedy Institute at Harvard and the whole operation was paid for by foundations.

The ecology movement is backed by enthusiastic environmental advertisers which include most of the major polluters. According to *Advertising Age,* the trade paper, companies rushing to buy prime time for Earth Day were Procter & Gamble, General Electric, Goodrich, DuPont, Standard Oil of New Jersey, International Paper, Phillips Petroleum, Coca-Cola, Chevron Oil, General Motors and Atlantic Richfield.

The new pollution-control industry looks to 20 percent growth annually over the next five years and counts on a $25 billion market. But the war and the depressed

economy is making the going slow. And the new industry
will have to knock out the old pollution industry before
it can begin to flourish. The old industry includes the
engineering and construction companies which make
their profits from building standard-sewage-treatment
works. They are supported within and without government by the sanitary and civil engineers, who are
organized through the Water Pollution Control Federa-
tion, a Washington professional society which also is a
lobby. In addition, the public-health doctors have always
lined up behind the sewermen, and most important, they
are solidly entrenched with the Public Works Committees of the Congress. Both public-works committees
work hand-in-glove with the Army Corps of Engineers;
it reflects the interests of the engineering and construction companies. The Public Works Committees are also
responsive to the big-city machines, which still consider
sewer contracts important for maintaining power and
providing employment.

CONCLUSIONS

The ecology movement was remote, separate and cut off
from the revolutionary surge sweeping through American society in the spring of 1970. A planned event, it had
no bearing on the war, political repression, blacks, the
poor, or any other factor which created the currents of
stress in the society. At last, "pollution control," "ecology," "environmental quality," whatever slogan it went
by, were no more than they were in Chadwick's time,

plans for advancing capitalism. Stripped of its façade, Environmental Action could offer no more than any other new industry, whose growth was tied to increasing pollution.

The most important single step toward alleviating pollution would be to change the national fuels policy, that collection of economic incentives which encourage the petroleum industry to search for oil. A beginning could be made by eliminating the depletion allowance, removing intangible drilling benefits, and denying US corporations tax deductions for foreign operations. It would be essential to abolish the import quotas, which create an artificial market for oil in the United States.

The government should attempt to manipulate the fuels markets to reduce pollution, even where that may not appear to be efficient. Thus, it should surely stop drilling for oil and gas on the outer continental shelf. The hazards from oil spills clearly outweigh the worth of this program. The vast deposits of oil and gas on the shelf should be developed in accordance with a coherent public policy, and only after the government conducts its own geological studies.

Congress should abolish the mining law of 1872 and establish a leasing system for mining minerals now covered by that law. Leases should be made only on the basis of sealed bids, with the government first establishing a minimum price. The bonus-bid scheme, where companies pay millions in cash payments to secure the leases, should also be abolished. The bonus-bid plan restricts the business to only the large companies, thereby increasing their monopoly. Leases should be granted for as long as the company is producing. No company should be permitted to sit on public land for more than one year. That would allow a decent interval for exploration. The US government, not the state-regulatory commissions, should set royalty fees and production levels on the outer continental shelf. In that way the governmen can regulate to some extent the flow of oil and natural gas to markets. And no company should be granted any lease to Federal lands unless the explorer can provide a detailed plan setting forth how the fuel will be mined, schemes for transporting it to markets, and its uses. No lease should be granted unless every participant in the process of handling the fuel, from the source through use, can demonstate that basic pollution standards will be met. If any party to such an agreement failed to abide by the plan, then all parties would be held absolutely liable for all damages, including those to any third party.

The Federal government should establish basic levels for pollutants—such as sulphur—in different fuels, and require industry to produce only those fuels. If the industry failed to do so, then the government should be equipped with injunctive powers to shut it down.

In general, the development, production and end use of fuels should be controlled through public policies. The fuels business, including electric utilities, should not be carried out for profit. One step in breaking the power of the fuels trusts would be to require petroleum companies to divest themselves of their chemical, coal and uranium subsidiaries, and to block their moves into consumer-goods business, such as plastics.

Air pollution from motor vehicles can obviously be dealt with in only one way: by developing another kind of engine. The prospects for doing this, through either the steam or gas-turbine engine, are considerably more promising than, for example, development of the SST. The government will have to finance an immense research and development program. At the same time, it will need to plan and construct mass-transit systems in cities.

Industries are largely responsible for polluting the water. The government should be equipped with injunctive powers to require pollution abatement or shut them down. Efforts by business to pass along the additional costs to consumers can be restricted by strict price regulations. The money to clean up pollution should come out of industrial profits. In many parts of the country, it is still possible to develop sewage systems in imitation of what Chadwick proposed long ago and along the lines of the recent Penn State experiments. That means planning a water supply and sewage system based on geographical areas, river basins and watersheds.

The proposals sketched out above are not meant as technical adjustments to existing governmental systems, "reforms" for controlling pollution. What they represent are different ways of attacking concentrated corporate power, the source of pollution, thereby opening up the possibilities of revolutionary change, and for reorganizing society and communities on different principles, with much more emphasis on geography. It is impossible to do much of anything about pollution without first achieving some sort of fundamental idea of community and a political economy.

Notes

Chapter 2

An "Inquiry into the Sanitary Condition of the Labouring Population of Great Britain" was published as a Report from the Poor Law Commissioners to the Home Department in London in 1842. While the conclusions have been summarized in numerous books and articles, the report itself is extraordinary, and much of this chapter is based on it. The description of conditions at Tranent is on p. 68; Dr. Duncan's report on Liverpool, p. 30; Death Statistics in Liverpool, p. 159; Chadwick's description of Edinburgh, p. 48, and the Villerme observation, p. 91. Biographies of Chadwick include *The Life and Times of Sir Edwin Chadwick*, by S. E. Finer, London, 1952; and *Edwin Chadwick and the Public Health Movement*, by Richard A. Lewis, London,

1952. Progress in sanitary reform is summarized in the testimony of A. D. Adrian in the interim report of the Royal Commission on Sewage Disposal, London, 1901. The early history of sewage in the United States in chronicled in various reports of the Massachusetts State Board of Health. See "Water Supply and Sewerage, Report of the State Board of Health of Massachusetts," January, 1888. For an account of the pollution in New York Harbor see the report, "Metropolitan Sewerage Commission," 1910. A good, general history of sewage practice is contained in *American Sewerage Practice*, by Leonard Metcalf and Harrison P. Eddy, New York, 1935.

Chapter 3

See *Conservation and the Gospel of Efficiency*, by Samuel P. Hays, Cambridge, 1959, p. 126. Lonergan's argument is summarized in "Stream Pollution and Purification," report of Senator Augustine Lonergan, January 30, 1935, published as Senate document No. 16. Karl Mundt's views are contained on p. 1046 of the Congressional Record for February 23, 1940. Also see "A Report on Water Pollution in the United States," the third report of the special advisory committee on water pollution of the National Resources Committee, February 16, 1939. House document No. 155, 76th Congress, 1st session. This report includes statistics on pollution along with the Roosevelt Administration's basic position. Reports of various conferences on the Raritan Bay, published by the Federal Water Quality Administration, provide basic data on pollution in the Bay.

Chapter 4

For general information on sewage treatment see Metcalf and Eddy, "American Sewerage Practice," *Fortune*, February 1970; included is an article summarizing modern sewage-treatment methods. For a more detailed approach see *Managing Water Quality: Economics, Technology, Institutions,* by Allen V. Kneese and Blair T. Bower, Baltimore, 1968. For information on management of sewage works see "Operation and Maintenance of Municipal Waste Treatment Plants," prepared by the General Accounting Office and published by the Senate Public Works Committee, November 1969. The Kneese book contains a description of the Delaware River Basin scheme. The Penn State experiment is described by Louis T. Kardos in *Environment,* March 1970. The description and analysis of hidden-sewer costs is taken from Francis Xavier Tannian's unpublished dissertation, "Water and Sewer Supply Decisions: A Case Study of the Washington Suburban Sanitary Commission," 1965. For details, write Tannian at the Division of Urban Affairs, University of Delaware, Newark, Delaware. He also has prepared similar studies of the Delaware River area. The best single source of material on the US water-pollution program is the General Accounting Office's "Examination into the Effectiveness of the Construction Grant Program for Abating, Controlling and Preventing Water Pollution," November 1969. It is available from the Senate Public Works Committee.

Chapter 5

For a current analysis of the petroleum industry see Senate anti-trust subcommittee hearings, The Petroleum Industry, March–May, 1969. *The Politics of Oil,* by Robert Engler, New York, 1961, provides sound basic material on the oil industry and its history. Also see *The Atlantic Monthly,* September 1969, for Ronnie Dugger's article, "Oil and Politics."

See the "Black Tide," *Environment* Magazine November 1969, for a general description of the oil-pollution problem.

The January 1970 *Ocean Industry* reports on the growth of
the tanker fleet. Sun Oil Company, Philadelphia, Pa., publishes an annual *Analysis of the World Tank Ship Fleet,*
which shows basic ownership patterns. The quotation from
the British experts and subsequent description of the
Torrey Canyon accident are from the *Torrey Canyon* Report of the Committee of Scientists on the "Scientific and
Technological Aspects of the *Torrey Canyon* Disaster,"
London 1967. Max Blumer's report "Oil Pollution of the
Ocean" is printed in Part 4, Water Pollution, hearings of the
Senate Subcommittee on Air and Water Pollution, 1969,
p. 1485. The list of oil spills is compiled from a record kept
by the Water Quality Administration. Parts 1 and 4 of the
1969 water-pollution hearings conducted by the Senate Subcommittee on Air and Water Pollution contain considerable
testimony on liability. For the dangers of shipping natural
gas see "Hazards of LNG Spillage" in "Marine Transportation," a report prepared by the Bureau of Mines, February
1970. For more general information on the various problems
posed by oil pollution see "Oil Pollution," a report to the
President by the Secretaries of Interior and Transportation,
February 1968.

There is little reliable information on the move to monopolize the fuels industry. But some of the best material has
been prepared by Dr. Bruce Netschert and his associates at
the National Economics and Research Associates, Inc. See
their "Competition in the Energy Markets," prepared for
the Senate anti-trust subcommittee, May 1970. The *NACLA
Newsletter,* published by the North American Congress on
Latin America, published three thorough articles on the
Hanna Industrial Complex. They appeared in the news,
letter from May to October 1968. For copies write to
NACLA, 57 Cathedral Station, New York, N.Y. 10025. A
list of companies with coal leaseholds on public lands is
available for inspection at the office of the Bureau of Land

Management, Interior Department. The description of leasing procedures is based on interviews with various officials. Lear's interview appeared in *The New York Times*, November 16, 1969.

For details on natural gas see Natural Gas Supply hearings, Minerals Subcommittee, Senate Interior Committee, November 13–14, 1969. See especially testimony of Nassikas, p. 115 and Netschert, p. 78. Also statement of Edward Berlin, p. 192 on the alleged gas shortage.

A detailed report on air pollution is available as "Vanishing Air," Grossman, 1970. This is the report by Nader's Raiders, edited by John Esposito. Also see "The Search for a Low-Emission Vehicle," staff report of the Senate Commerce Committee, Washington, 1969. Loesch was quoted in the Grand Junction, Colo., *Daily Sentinel*, February 1, 1970.

There are two thorough studies of the Santa Barbara oil spill: *The Santa Barbara Oil Spill*, by Malcolm F. Baldwin, published by the Conservation Foundation, provides a thorough study of the oil spill at Santa Barbara. Also see *Geology, Petroleum Development and Seismicity of the Santa Barbara Channel Region*, published by the US Geological Survey as Professional Paper #679. Especially see "Geological Characteristics of the Dos Cuadra Offshore Oil Field," by T. H. McCulloh, p. 29. Also *Review of the Santa Barbara Channel Oil Pollution Incident*, published by the Water Quality Administration, July 1969; and Part 3, water-pollution hearings of the Senate subcommittee on air and water pollution, February 24–25, 1969.

Chapter 6

Secretary Hickel's compliments to US Steel were contained in an Interior department release dated April 30, 1969. Jack B. Schmetterer, first Assistant US Attorney in Chicago, delivered his attack on the Federal Water Quality Administration in an April 1970 speech entitled, "Federal Enforcement of Pollution Laws."

See *The Population Bomb,* by Paul Ehrlich, New York, 1968; the "Nonsense Explosion" by Ben Wattenberg in *The New Republic,* April 4, 1970. Also *Road to Survival* by William Vogt, New York, 1948. The quotations are on pp, 287, and 282 respectively; *Off Our Backs,* the Washington Women's Liberation newspaper for March 19, 1970, p. 5. See "The Social Context of US Population Control Programs in the Third World," by William Barclay, Joseph Enright and Reid T. Reynolds. This paper was presented to the annual meeting of the Population Association of America, April 1970. To obtain a copy write Reynolds, c/o Department of Sociology, Cornell University, Ithaca, New York. "Why the Population Bomb Is a Rockefeller Baby," by Steve Weissman in the May 1970 *Ramparts* provides a good summary and analysis of the political development of the population lobby.

"The Making of a Pollution-Industrial Complex," by Martin Gellen, also in the May 1970 *Ramparts,* provides an account of the new industry.

Index

ABOUT THE AUTHOR

JAMES RIDGEWAY, 33, was born in Auburn, New York. He is a contributing editor of the *New Republic,* and an editor and founder of *Hard Times,* the radical weekly paper. Mr. Ridgeway is the author of *The Closed Corporation,* a report on the university-industrial complex. He is married and lives in Washington, D.C.